The JCT Intermediate Building Contracts 2005

Third Edition

David Chappell

Editorial offices:
Blackwell Publishing Ltd, 9600 Garsington Road, Oxford OX4 2DQ, UK
Tel: +44 (0)1865 776868
Blackwell Publishing Inc., 350 Main Street, Malden, MA 02148-5020, USA
Tel: +1 781 388 8250
Blackwell Publishing Asia Pty Ltd, 550 Swanston Street, Carlton, Victoria 3053, Australia
Tel: +61 (0)3 8359 1011

First edition published by Legal Studies and Services (Publishing) Ltd 1991
Second edition published by Blackwell Science 1999
Reprinted 2002
Third edition published by Blackwell Publishing 2006

ISBN-10: 1-4051-4049-6
ISBN-13: 978-1-4051-4049-2

Library of Congress Cataloging-in-Publication Data

Chappell, David.
The JCT intermediate building contracts 2005/
David Chappell–3rd ed.
p. cm.
Rev. ed. of: JCT intermediate form of contract.1999.
Includes index.
ISBN-13: 978-1-4051-4049-2 (hardback : alk. paper)
ISBN-10: 1-4051-4049-6 (hardback : alk. paper)
1. Construction contracts–Great Britain.
I. Chappell, David. JCT intermediate form of contract. II. Title.

KD1641.C486 2006
343.41'07869–dc22

2006002367

A catalogue record for this title is available from the British Library

Set in 10.5/12.5pt Palatino
by Graphicraft Limited, Hong Kong
Printed and bound in Great Britain
by TJ International Ltd, Padstow, Cornwall

The publisher's policy is to use permanent paper from mills that operate a sustainable forestry policy, and which has been manufactured from pulp processed using acid-free and elementary chlorine-free practices. Furthermore, the publisher ensures that the text paper and cover board used have met acceptable environmental accreditation standards.

For further information on Blackwell Publishing, visit our website:
www.blackwellpublishing.com

Contents

Preface to the Third Edition

I hope that this will be a useful book for architects, quantity surveyors, project managers and contractors. The book explains not only what the contract provisions mean, but also their practical applications. It is important to say that it is not a clause-by-clause analysis. Instead, the form has been considered under topic headings, which enables all the clauses relating to a particular topic to be dealt with together. The topics cover the roles of the parties and important matters such as payment, claims, design issues, determination and dispute resolution. Legal language has been avoided in favour of simple explanations of legal concepts, and the text is supported by flow-charts, tables and sample letters. Where possible, a clear statement of the legal position in a variety of common circumstances has been stated, together with references to relevant decided cases so that those interested can read further.

Occasionally, where the precise legal position is not clear, a view has been offered. This book is intended to be a practical working tool for all those using IC and ICD.

Much has changed since the last edition was written. JCT completely revised the Intermediate Building Contract in 2005. The structure of the contract changed dramatically with the introduction of contract particulars, warranties and schedules. Clauses have been rearranged, re-numbered and re-worded, sectional completion has been incorporated, and some terminology has been changed. An alternative version includes provision for contractor design. This edition has been thoroughly revised to deal with all these important changes, including a separate chapter dealing with the contractor's designed portion. Almost 50 additional cases have been included, and all relevant legislation.

Throughout the text the contractor has been referred to as 'it' on the basis that it is a corporate body.

The first edition of this book was written with the late Professor Vincent Powell-Smith, an eminent commentator in this field.

David Chappell
Wakefield
January 2006

CHAPTER ONE
THE PURPOSE AND USE OF IC AND ICD

1.1 *The background*

The JCT Intermediate Form of Building Contract (IFC 84) was first published in September 1984 as a member of the growing family of standard contract forms issued by the Joint Contracts Tribunal. IFC 84 was amended 12 times after its issue, the latest amendment dated April 1998. At the end of 1998 IFC 84 was reprinted, complete with all 12 amendments, and further corrected to deal with an accumulation of inconsistencies and inaccuracies, particularly in amendment 12. The new edition was referred to as IFC 98. In May 2005 a new version of the form was produced, now called the Intermediate Building Contract (IC). A separate edition (ICD) was issued to cater for instances where it was desired to have the contractor carry out part of the design.

It is intended that IC should fill the gap between the JCT Minor Works Building Contract (MW), which is suited to smaller projects, and the very complex JCT Standard Building Contract (SBC). Table 1.1 sets out the JCT forms currently available. The most significant features of IC are as follows:

- The provisions (clause 3.7 and schedule 2) enabling the architect to select specialist sub-contractors by 'naming' them
- The provision (clause 3.15) for the architect to issue instructions regarding failure of work and the inspection of similar work.

In arrangement, the form follows the pattern of SBC in general structure and in setting out the conditions under nine main headings. IC is arranged as follows:

ARTICLES OF AGREEMENT
Contract particulars

Table 1.1
JCT forms of contract in current use

Title	Contract documents	Comments
Standard Building Contract (SBC)	Drawings, bills of quantities (*variants*: with approximate quantities *or* specification *or* activity schedule)	Complex and sophisticated: for use on substantial traditional projects where there is a requirement for tight and detailed drafting
Intermediate Building Contract (IC)	Drawings and *either* specification *or* work schedules *or* bills of quantities	Suitable on traditional projects where the work is not long or complex, without complex service installations. Value from about £120,000 to £400,000. There is a variant (ICD) allowing for contractor's designed portion
Minor Works Building Contract (MW)	Drawings and *either* specification *or* work schedules	For traditional works of simple character and short duration. Value up to about £120,000. There is a variant (MWD) allowing for contractor's designed portion
Prime Cost Building Contract (PCC)	Specification with *or* without drawings	Useful for early start on site and when not possible to prepare complete design information before commencement. Little used in practice – prime cost plus percentage fee
Design and Build Contract (DB)	Employer's Requirements, Contractor's Proposals and Contract Sum Analysis	If contractor is to be responsible for all the design
Management Building Contract (MC)	Drawings, specification, cost plan and schedules	Low risk to the management contractor. Work carried out by works contractors
Construction Management Agreement (CM/A)	Drawings, specification and schedules	A construction manager carries out management functions. Work carried out under separate trade contracts
Major Project Construction Contract (MP)	Flexible arrangements	Used for major projects where the employer is used to procuring large projects and has its own detailed procedures
Measured Term Contract (MTC)	Schedule of rates	For use by employers who need regular maintenance and minor works to be carried out by one contractor over a specified period on receipt of individual orders

CONDITIONS

(1) Definitions and interpretation
(2) Carrying out the Works
(3) Control of the Works
(4) Payment
(5) Variations
(6) Injury, damage and insurance
(7) Assignment and collateral warranties
(8) Termination
(9) Settlement of disputes

SCHEDULES

(1) Insurance options
(2) Named sub-contractors
(3) Forms of bonds
(4) Fluctuations option – contribution, levy and tax fluctuations

In this book this sequence has not been followed. Instead, an arrangement has been adopted which allows for topic grouping.

1.2 *IC documentation*

In addition to the printed contract form itself, which must be completed and executed formally by the employer and the contractor, there is also a set of supporting documentation:

- Intermediate Sub-Contract Agreement (ICSub/A)
- Intermediate Sub-Contract Conditions (ICSub/C)
- Intermediate Sub-Contract with Sub-Contractor's Design Agreement (ICSub/D/A)
- Intermediate Sub-Contract with Sub-Contractor's Design Conditions (ICSub/D/C)
- Tender and Agreement for Named Sub-Contractors under IC (ICSub/NAM) – summarised in Appendix A
- Sub-Contract Conditions for Named Sub-Contractors under IC (ICSub/NAM/C) – summarised in Appendix B.

The Joint Contracts Tribunal has also published two related guides:

- Intermediate Building Contract Guide (IC/G)
- Intermediate Sub-Contract Guide (ICSub/G).

The JCT has also prepared a form of Intermediate Named Sub-Contractor/Employer Agreement under IC (ICSub/NAM/E). This is summarised in Appendix C. It is a vital document. The architect is responsible for arranging for its completion on the employer's behalf. Its effect is to create a direct contractual relationship between the employer and the named sub-contractor for limited purposes, and it gives the employer the necessary protection against design and related failures in a named sub-contractor's work.

1.3 The use of IC

IC is issued for contracts in the range between SBC with quantities and MW. It is suitable for use where:

- The proposed Works are simple in content, involving the normally recognised basic trades and skills of the industry
- The proposed Works are without any building service installations of a complex nature; and adequately specified, or specified and billed as appropriate before the issue of tenders
- A contract administrator and quantity surveyor administer the conditions.

IC is suitable for use only where the criteria are met. It may be deduced that IC is suitable for most such projects, with or without quantities, whose value is between £90,000 and £380,000 (at 1998 prices), with a contract period of not more than 12 months.

Contract value is probably not the deciding factor, and IC may be suitable for somewhat larger contracts provided that the three basic criteria are met. Four things would preclude the use of the form:

- Uncertainty of work
- Complexity of work
- The wish or need for nominated sub-contractors and nominated suppliers
- The wish to make the contractor responsible for the whole of the design.

In summary, IC is highly suitable for use on projects of medium size with a simple work content. IC should not be used merely to avoid the complexities of SBC. In their own way the IC provisions are just as complicated.

Provision for the Works to be done in sections is included in the wording of the IC and ICD, so that there is no necessity for the use of a separate supplement.

Table 1.2 sets out the main differences between the third edition of the Association of Consultation Architects' Form of Building Agreement 1998 (Revised 2003) (ACA 3), SBC and IC.

1.4 Completing the form

Normally, a formal contract will be executed by the parties, and the printed form IC will be used for this purpose. There are some differences when ICD is used to reflect the inclusion of provision for contractor design in the ICD form.

The articles of agreement

Page 1 should be completed with the descriptions and addresses of the employer and the contractor. It is important to include the registration number if one or both are limited companies, because company names may change, but the number identifies the company. The author has experience of dealing with two companies which effectively swapped names – resulting in a contractor which had no assets. The date will not be inserted until the form is signed or executed as a deed by the parties.

The first recital must be completed with special care, because the description of 'the Works' is important in many respects, not least when considering the question of variations. The contract drawings are to be listed in the second recital unless they are too numerous, in which case reference should be made to where the list can be found, and it should be attached to the contract documents. Where ICD is used, the list of drawings is in the third recital. The second recital is used to state the nature and extent of work which will be part of the contractor's designed portion. This must not include any work which is to be carried out by a named sub-contractor.

In IC, the appropriate deletions should be made to both the third and fourth recitals as indicated in the footnotes (these are the fourth and fifth recitals in ICD). Consistency is important, because the contractor is required to carry out and complete the works in accordance with the contract documents as identified in the definitions. Reference to the priced activity schedule in the fourth recital (the fifth recital in ICD) should be deleted if it is not to be used.

Table 1.2
ACA 3, SBC and IC compared

Subject	ACA 3	SBC	IC
Cost limits (upper)	None	None, but can be used for smaller projects with specialist work content or complex services	£400,000 and Works must be of simple content and not long duration
Contract documents	Drawings, schedule of rates, bills of quantities, specification	Drawings, work schedules, bills of quantities, specification, activity schedules	Drawings, work schedules, bills of quantities, specification
Date for possession	To be given to contractor as specified in time schedule. Provision for possession to be given in parts	Must be given on the date stated in the contract particulars; power to defer possession for up to six weeks	As SBC
Extension of time	Alternative clauses – 11.5.1 covers failure to comply with CDM Regulations and act, instruction, default or omission on the part of the employer, etc. Clause 11.5.2 similar to SBC. Detailed notification provisions	Detailed provisions and a long list of grounds; comprehensive	Similar to SBC
Liquidated damages	Alternative clauses – conventional liquidated damages *or* unliquidated damages. Architect's certificate that the Works are not fit and ready for taking over is precondition to deduction by the employer	Liquidated damages deductible by employer if architect gives certificate of non-completion. Provision for adjustment if later extension issued	Similar to SBC

Table 1.2 *Cont'd*

Subject	ACA 3	SBC	IC
Variations	Detailed provisions – provision for the submission of detailed estimates by the contractor including loss and/or expense. Detailed valuation rules	Detailed provisions with valuation rules, provision for quotation requested by the architect	Similar to SBC, but no provision for architect to request a quotation
Payment and retention	Interim payments monthly, but conditional on contractor making application 95%. Alternative stage payments. Final account must be submitted by contractor within 30 days of expiry of the maintenance period	Interim payments monthly. Documents for final account to be submitted by the contractor within 6 months of issue of practical completion certificate. Retention released at practical completion and completion of making good	Similar to SBC. Retention release at practical completion and final certificate
Fluctuations	Optional fluctuations clause based on ACA Index – 80% only payable	Alternative provisions for contributions, levy and tax fluctuations, or labour and materials and tax fluctuations, or formula adjustment where there are contract bills	As SBC except only the contribution levy and tax fluctuation option
Loss and/or expense	Detailed provisions entitling contractor to claim for disturbance of regular progress caused by employer's or architect's acts,	List of relevant matters and procedures; very comprehensive	Similar to, but less detailed than, SBC

Table 1.2 *Cont'd*

Subject	**ACA 3**	**SBC**	**IC**
	omissions, default, or negligence. Notice of claim required if payment to be included in interim certificate and estimates of cost required. Otherwise architect adjusts in final certificate, but contractor loses interest element		
Selection of sub-contractors	Provision for named sub-contractors and suppliers. Architect can be involved in negotiations. Contractor fully responsible for performance including design	List of three selected. No longer any provision for nominated sub-contractors or suppliers	Sub-contractors can be named in contract documents or in instructions. No provision for naming suppliers
Dispute resolution	Alternative methods: • Conciliation and • Adjudication (CIC Procedure) and *either* • Litigation *or* • Arbitration	Alternative methods: • Mediation and • Adjudication (The Scheme) and *either* • Arbitration *or* • Litigation	As SBC
Advantages/ disadvantages	Flexible, simple payment scheme, provision for design responsibility by contractor with indemnity cover. Useful range of alternative clauses. Time periods may not be realistic	Comprehensive, widely known, very detailed and complex	Relatively complex, widely used

An information release schedule is rarely provided by the contractor, for the very good reasons set out in Chapter 5, and therefore the sixth recital (the ninth recital in ICD) is often deleted. The appropriate deletions must be made in the eighth recital (the eleventh recital in ICD).

ICD has two additional recitals, the sixth and eleventh, which do not require any insertions or deletions.

The amount of the contract sum must be inserted in words and figures in article 2. It is important to remember that the contract sum will always be this figure. If the sum needs to be adjusted by the addition of the value of variations or loss and/or expense, it will not be a new contract sum, but merely the 'adjusted contract sum'.

The name of the architect or, if not registered in accordance with the Architects Act 1997, the contract administrator is to be inserted in article 3, and the name of the quantity surveyor in article 4. Article 5 is for the name of the planning supervisor if it is not to be the architect, and the appropriate deletion should be made. A similar situation exists in article 6 if the principal contractor is not to be the contractor.

Articles 7, 8 and 9 deal with the dispute resolution procedures.

Contract particulars

This replaces the appendix in IFC 98. It should be fully and carefully completed by the architect, consistent with the information given to the contractor at tender stage, because the tender is based on the information supplied. Unlike the appendix in IFC 98, the contract particulars are divided into parts 1 and 2 dealing with general matters (such as dates for possession and completion) and collateral warranties. The latter part is newly introduced in this contract.

Entries which are inconsistent with the printed conditions generate much income for lawyers. Note that each set of contract particulars will be completed in a unique way to suit the particular project, the circumstances, and the employer's requirements.

Attestation

Alternative attestation clauses are provided, as the contract may be executed by hand or as a deed. The employer will have made this decision at pre-tender stage, and there is an important practical difference.

The Limitation Act 1980 specifies a limitation period – the time within which an action may be commenced – of six years where the contract is merely signed by the parties, or 12 years in the case of a contract entered into as a deed. These periods begin to run when the cause of action accrues, which, in the case of a contractual claim, is the date of the breach of contract. It is not always easy to fix this date, although it may be thought that it must be the date that the defective work was done. In construction contracts the limitation period is usually taken as being on the date of practical completion: *Borough Council of South Tyneside* v. *John Mowlem & Co, Stent Foundations Ltd and Solocompact SA* (1997); *Tameside Metropolitan Borough Council* v. *Barlows Securities Group Services Ltd* (1991). From the employer's point of view, it is usually sensible to contract as a deed.

In order to contract by deed, it used to be necessary to affix red wafer seals, or to write 'LS' (meaning 'in the place of the seal') if the seals were not obtainable. Provisions in the Law of Property (Miscellaneous Provisions) Act 1989 applicable to individuals and the Company Act 1989 applicable to companies have changed the situation by abolishing the necessity to seal a deed. In the case of companies, it must be stated on the face of the document that it is a deed, and it must be signed by two directors or a director and a company secretary. In Northern Ireland the need for a seal in the case of companies was removed by the Companies (No. 2) Order (Northern Ireland) 1990, but currently the requirement for a seal in the case of individuals still remains. That position is expected to change.

If, as is not generally desirable, any amendments are made to the printed text, or any clauses are deleted, these should be initialled by both contracting parties.

CHAPTER TWO
CONTRACTS COMPARED

IC is a very flexible contract, and its content establishes it as a member of the Joint Contracts Tribunal family. Its provisions closely parallel those of SBC in many respects, including the format and layout. It is silent on some matters. Such gaps would have to be filled in by the common law.

Table 2.1 compares IC with SBC and MW, and also with the third edition of the ACA Form of Building Agreement, published in 1998, as revised in 2003 (ACA 3).

Table 2.1
IC clauses compared with those of other common standard contracts

IC clause	Description	SBC clause	MW clause	ACA 3 clause	Comment (on IC unless otherwise stated)
1	**Definitions and interpretations**				
1.1	Definitions	1.1	1.1	–	
1.2	Reference to clauses, etc.	1.2	–	–	
1.3	Agreement to be read as a whole	1.3	1.2	1.3	The printed conditions, etc. prevail over any specially prepared clauses if there is a conflict
1.4	Headings, references to persons, legislation, etc.	1.4	1.3	23.2	
1.5	Reckoning periods of days	1.5	1.4	Recital G	
1.6	Contracts (Rights of Third Parties) Act 1999	1.6	1.5	Recital J	
1.7	Giving or service of notices	1.7	1.6	23.1	
1.8	Electronic communications	1.8	–	–	
1.9	Issue of architect's certificates	1.9	–	23.1	
1.10	Effect of final certificate	1.10	–	19.5	
1.11	Effects of certificates other than final certificate	1.11	–	19.5	
1.12	Applicable law	1.12	1.7	25.11	

Table 2.1 *Cont'd*					
IC clause	**Description**	**SBC clause**	**MW clause**	**ACA 3 clause**	**Comment (on IC unless otherwise stated)**
2	**Carrying out the Works**				
2.1	General obligations	2.1	2.1	1.1–1.2	
2.2	Materials, goods and workmanship	2.3	2.1	–	
2.3	Fees and charges	2.21	2.6	–	
2.4	Date of possession	2.4	2.2	11.1	
2.5	Deferment of possession	2.5	–	–	ACA 3 empowers architect to order acceleration or postponement (11.8)
2.6	Early use by employer	2.6	–	–	
2.7	Work not forming part of the contract	2.7	–	10.1–10.4	
2.8	Contract documents	2.8	–	2.1	
2.9	Levels and setting out	2.10	–	–	
2.10	Construction information	2.9	–	2.1	
2.11	Further drawings, details and instructions	2.12	2.3	2.1	
2.12	Bills of quantities	2.13	–	1.4	
2.13	Instructions on errors, omissions and inconsistencies	2.14	2.4	1.4–1.5	
2.14	Instructions – additions to contract sum, exceptions	2.14.3	–	–	
2.15	Divergences from statutory requirements	2.17	2.5	1.6	

Table 2.1 *Cont'd*

IC clause	Description	SBC clause	MW clause	ACA 3 clause	Comment (on IC unless otherwise stated)
2.16	Emergency compliance with statutory requirements	2.18	–	–	
2.17	Materials and goods – on site	2.24	–	6.1	It is doubtful that these provisions would be effective against a retention of title clause in a supplier's contract of sale
2.18	Materials and goods – off site	2.25	–	6.1	It is doubtful that these provisions would be effective against a retention of title clause in a supplier's contract of sale
2.19	Notice of delay – extensions	2.27–2.28	2.7	11.6–11.7	
2.20	Relevant events	2.29	–	11.5	
2.21	Practical completion and certificates	2.30	2.9	12.1	In ACA 3, the concept of 'taking over' is similar to practical completion
2.22	Certificate of non-completion	2.31	–	11.2	
2.23	Liquidated damages	2.32	2.8	11.3	
2.24	Repayment of liquidated damages	2.32.3	–	11.4	
2.25	Contractor's consent to partial possession	2.33	–	13.1	
2.26	Practical completion date	2.34	–	13.2	

Table 2.1 *Cont'd*

IC clause	Description	SBC clause	MW clause	ACA 3 clause	Comment (on IC unless otherwise stated)
2.27	Defects, etc.	2.35	–	–	
2.28	Insurance	2.36	–	–	
2.29	Liquidated damages	2.37	–	–	
2.30	Rectification	2.38	2.10	12.2–12.3	
2.31	Certificate of making good	2.39	2.11	–	
3	**Control of the Works**				
3.1	Access for architect	3.1	–	4.1–4.2	
3.2	Person-in-charge	3.2	3.2	5.2–5.3	
3.3	Clerk of works	3.4	–	–	
3.4	Replacement of architect/quantity surveyor	3.5	Article 3	Recital E	
3.5	Consent to sub-letting	3.7	3.3	9.2	
3.6	Conditions of sub-letting	3.9	3.3	–	
3.7	Named sub-contractors	–	–	9.3–9.7	
3.8	Compliance with instructions	3.10	3.4	8.1	No provision for confirmation of oral instructions
3.9	Non-compliance with instructions	3.11	3.5	–	
3.10	Provisions empowering instructions	3.13	–	–	
3.11	Instructions requiring variations	3.14	3.6.1	8.2	
3.12	Postponement of work	3.15	–	11.8	

Table 2.1 *Cont'd*

IC clause	Description	SBC clause	MW clause	ACA 3 clause	Comment (on IC unless otherwise stated)
3.13	Instructions on provisional sums	3.16	3.7	–	
3.14	Inspection – tests	3.17	–	8.1	
3.15	Work not in accordance with the contract	3.18	–	8.1	
3.16	Instructions as to removal of work	3.18.1	–	8.1	
3.17	Exclusion of persons from the Works	3.21	3.8	8.1	
3.18	Undertakings to comply with CDM Regulations	3.25	3.9	26	
3.19	Appointment of successors	3.26	3.10	26.4	
4	**Payment**				
4.1	Work included in contract sum	4.1	–	–	
4.2	Adjustment only under the conditions	4.2	–	15.1	
4.3	VAT	4.6	4.1	16.8	
4.4	CIS	4.7	4.2	24	
4.5	Advance payment	4.8	–	–	
4.6	Interim certificates and valuations	4.9, 4.11 and 4.12	4.3	16.1	
4.7	Amounts due on interim certificates	4.10	4.3	16.2	
4.8	Interim certificates – payment	4.13	4.6	16.3 & 16.6	

Table 2.1 *Cont'd*

IC clause	Description	SBC clause	MW clause	ACA 3 clause	Comment (on IC unless otherwise stated)
4.9	Interim payment on practical completion	–	4.5	–	
4.10	Interest on percentage withheld	4.18.1	4.4	16.4–16.5	
4.11	Contractor's right of suspension	4.14	4.7	–	This is a right under section 112 of the Housing Grants, Construction and Regeneration Act 1996
4.12	Off-site materials and goods	4.17	–	16.2	
4.13	Adjustment of contract sum	4.5	–	19.1	
4.14	Issue of final certificate	4.15	4.8–4.9	19.2–19.3	
4.15	Fluctuations	4.21	4.10–4.11	18	Under ACA 3, the fluctuations clause is optional. A single index is used and only 80% is payable to the contractor
4.16	Fluctuations – named sub-contractors	–	–	–	
4.17	Disturbance of regular progress	4.23	–	7	
4.18	Relevant matters	4.24	–	7.1	
4.19	Reservation of contractor's rights	4.26	–	–	

Table 2.1 *Cont'd*

IC clause	Description	SBC clause	MW clause	ACA 3 clause	Comment (on IC unless otherwise stated)
5	**Variations**				
5.1	Definition of variations	5.1	–	–	
5.2	Valuation of variations and provisional sum work	5.2	3.6.3	16	
5.3	Measurable work	5.6	–	–	
5.4	Daywork	5.7	–	–	
5.5	Change of conditions for other work	5.9	–	–	
5.6	Additional provisions	5.10	–	–	
6	**Injury, damage and insurance**				
6.1	Liability of contractor – personal injury or death	6.1	5.1	6.3	
6.2	Liability of contractor – injury or damage to property	6.2	5.2	6.3	
6.3	Injury or damage to property – Works and site materials excluded	6.3	5.2	6.3	
6.4	Contractor's insurance of liability	6.4	5.3 & 5.5	6.3	
6.5	Contractor's insurance of liability of employer	6.5	–	6.5	
6.6	Excepted risks	6.6	–	–	
6.7	Works insurance – options	6.7	5.4 & 5.5	6.4	
6.8	Related definitions	6.8	–	–	

Table 2.1 *Cont'd*

IC clause	Description	SBC clause	MW clause	ACA 3 clause	Comment (on IC unless otherwise stated)
6.9	Sub-contractors – specified perils cover under AR policies	6.9	–	–	
6.10	Terrorism cover – non-availability	6.10	–	–	
6.11	Application of Joint Fire Code	6.13	–	–	
6.12	Compliance with Code	6.14	–	–	
6.13	Breach of Code	6.15	–	–	
6.14	Code – revisions	6.16	–	–	
7	**Assignment and collateral warranties**				
7.1	Assignment	7.1	3.1	9.1	
7.2	References	7.3	–	–	
7.3	Notices	7.4	–	–	
7.4	Execution of warranties	7.5	–	–	
7.5	Warranties – purchasers and tenants	7C	–	–	
7.6	Warranty – funder	7D	–	–	
7.7	Sub-contractors' warranties – purchaser /tenant/funder	7E	–	–	
7.8	Sub-contractors' warranties – employer	7F	–	–	
8	**Termination**				
8.1	Meaning of insolvency	8.1	6.1	–	
8.2	Notices under 8	8.2	6.2	–	

Table 2.1 *Cont'd*

IC clause	Description	SBC clause	MW clause	ACA 3 clause	Comment (on IC unless otherwise stated)
8.3	Other rights, reinstatement	8.3	6.3	22.5	
8.4	Employer termination – default by contractor	8.4	6.4	20.1	
8.5	Insolvency of contractor	8.5	6.5	20.3	
8.6	Corruption	8.6	6.6	–	
8.7	Consequences of termination by employer	8.7	6.7	22.1, 22.4, 22.6 & 22.7	
8.8	Employer's decision not to complete the Works	8.8	–	–	
8.9	Contractor termination – default by employer	8.9	6.8	20.2	
8.10	Insolvency of employer	8.10	6.9	20.3	
8.11	Either party termination	8.11	6.10	21	
8.12	Consequences of contractor or either party termination	8.12	6.11	22.2/ 22.3, 22.4, 22.6 & 22.7	
9	**Settlement of disputes**				
9.1	Mediation	9.1	7.1	–	
9.2	Adjudication	9.2	7.2	25B	
9.3–9.8	Arbitration	9.3–9.8	7.3	25C	
–	Conciliation	–	–	25A	

CHAPTER THREE
CONTRACT DOCUMENTS AND INSURANCE

3.1 *Contract documents*

3.1.1 Types and uses

IC is designed to cover a broad range of work and costs. To give effect to this intention, there are several possible combinations which can constitute the 'contract documents'.

What are contract documents? They are those documents which give legal effect to the intentions of the parties. In principle, the contract documents may consist of, and contain, whatever the parties wish. IC sets out the options in the fourth recital:

- The contract drawings and the specification, priced by the contractor; or
- The contract drawings and the work schedules, priced by the contractor; or
- The contract drawings and the bills of quantities, priced by the contractor; or
- The contract drawings and the specification and the contract sum analysis or schedule of rates.

One of these options, together with the agreement, the conditions annexed to the recitals and any relevant named sub-contractor information, forms the contract documents. They must all be signed by, or on behalf of, the parties.

The options must be studied carefully in order to arrive at the most suitable combination for a particular project.

The contract drawings and the specification priced by the contractor

This combination is usually appropriate for relatively small Works or for Works of a simple nature. It must be remembered that, if bills of quantities are not provided, the contractor will have to take off its own quantities in order to arrive at a tender sum. The prices eventually put to the specification will inevitably be somewhat rough and ready unless the specification is a model of clarity. As the priced specification will be the basis of the valuation of variations (clause 5.3), there could be pitfalls for contractor and employer alike.

The contract drawings and work schedules priced by the contractor

This combination is useful for work which is somewhat more complicated than the last example but does not warrant full bills of quantities. Again, the contractor will be obliged to take off its own quantities, and the employer, or the building industry in general, will in effect bear the cost. Work schedules demand that there is agreement on the order in which the work will be carried out. Even on a relatively simple job, the contractor may be able to devise cheaper and more efficient methods of arriving at the result than those envisaged by the architect. Some form of two-stage tendering procedure is indicated to allow such a contribution by the contractor before the schedule is completed. Whether this is warranted will depend on all the circumstances. A priced schedule is more comprehensible, for the purpose of valuations, than a priced specification, but it requires the utmost clarity of thought to prepare properly.

The contract drawings and the bills of quantities priced by the contractor

If the size of the job justifies it, this must be the most satisfactory all-round combination. It is tried and tested, and each party knows not only the price of the whole job, but also the cost of variations. Monthly valuations are simplified, and there is less likelihood that the contractor will perpetrate some terrible undetected mistake at tender stage.

The contract drawings and the specification and the contract sum analysis or schedule of rates

This combination requires the contractor to take off its own quantities and supply a total price, which will become the contract sum. The contractor is not required to price the specification but, instead, to supply the employer with a contract sum analysis of the stated sum or a schedule of rates on which the stated sum is based. The first point to note is that the contract does not include either the contract sum analysis or the schedule of rates as a contract document. This appears to be a serious miscalculation, although clearly it is intentional (see the definition of 'Contract Documents' in clause 1.1). There seems to be no good reason why either document should not be as important as, say, the priced specification in the first option. It is suggested that, if the employer uses option B of the fourth recital, the definition should also be amended to include the pricing documents as contract documents.

The contract sum analysis is stated (fourth recital) to mean an analysis of the contract sum in accordance with the stated requirements of the employer. The definition purposely leaves room for the architect to require the contractor to provide the analysis in any form required, presumably including bills of quantities.

The schedule of rates is more familiar, and may be thought more useful than the work schedules in another option. It is likely that this option will be used in most cases where bills of quantities are not prepared. A schedule of rates may be useful where the content of the work is not precisely known. This form is not designed for that situation, however, and the Prime Cost Building Contract would be indicated (PCC).

The contractor may also have provided the employer with a priced activity schedule. If provided, it will be used in calculating interim payments (clause 4.7.1.1). The activity schedule is usefully defined in footnote [6]. Essentially, it is a schedule of activities attached to the contract. Each activity is priced and, crucially, the sum of the prices must equal the contract sum less provisional sums, prime cost sums, the contractor's profit on such sums, and any work for which an approximate quantity is included in the contract documents. The use of such a schedule is said to give less scope for the quantity surveyor to undervalue work done.

The contract drawings are to be noted by number in the designated place in the first recital. Occasionally, there is some disagreement over the number and type of drawings to be designated contract drawings. They must be: sufficiently detailed to show the location and extent of the work; those from which the contractor obtained information to submit the tender; and related to the other contract documents.

Ideally, the contract drawings should include every drawing prepared for the work. In practice, this is not always possible, but if drawings and specification are being used to obtain tenders, the drawings must be detailed enough to allow the contractor to carry out its own taking off. If bills of quantities are provided, small details may be omitted from the drawings provided they are included in the bills of quantities.

The further drawings and details which the architect is to provide under clause 2.11 are not contract documents. If they show different or additional or less work and materials than are shown on the contract documents, the contractor will be entitled to a variation.

All the contract documents must be signed and dated by both parties. That means every separate drawing or separate piece of paper, but not every sheet of the specification or bills of quantities – signing the cover or the last page is sufficient. An endorsement on each document should read: 'This is one of the contract documents referred to in the agreement dated . . .' or other words to the same effect.

3.1.2 Importance and priority

The significance of the contract documents has already been discussed briefly. If a dispute arises, and it is necessary to discover what was agreed between the parties, the adjudicator, the arbitrator or the court will look at the contract documents.

In various places throughout the contract reference is made to work being or not being in accordance with the contract (e.g. clause 3.15). This means in accordance with what is contained in the contract documents. A problem arises if the documents are in conflict. Clause 4.1 sets out rules which are to be followed depending on which combination of documents has been chosen.

In order to decide on the quality and quantity of work agreed to be carried out for the contract sum, it is necessary to look at the particular documents.

Drawings and specification (or work schedules)

They must be read together, provided that no quantities are shown. If there is a conflict, the drawings are to be given preference over the specification or work schedules. If quantities are shown for some items, those quantities will prevail. So if the drawings show a total of 60 holding-down straps, but the specification states that 50 holding-down straps are required, the contractor may quite correctly price for 50 holding-down straps. If, at a later stage in the contract works, the architect decides that he or she does need 60 straps as shown on the drawings, the employer will have to pay for the additional 10.

If, on the other hand, the straps are mentioned in the specification without being quantified, the contractor will be deemed to have priced for 60 straps.

Drawings and bills of quantities

The quality and quantity of work shown in the bills of quantities will prevail.

Clause 4.1 will be welcomed by contractors, because it clarifies a situation which can be a source of dispute when the architect for a particular project maintains that the contractor should price for everything, whether it is mentioned in the specification or on the drawing. This position, however, still applies in this contract unless quantities are mentioned. In practice, the result is likely to be that the architect will have to take great care when he or she prepares the specification, or the employer will face large bills for additional work.

Clause 1.3 provides that nothing in the bills of quantities/the specification/the work schedules (as appropriate) will override or modify the agreement or the conditions. In effect, it means that nothing in the printed form can be amended by inserting a clause in the bills of quantities, etc. which attempts to modify or alter what any printed clause says. This is so even if the insertion is written in ink and signed by both parties. The clause has been upheld by the courts: *English Industrial Estates Corporation Ltd* v. *George Wimpey & Co Ltd* (1972). The way to amend the printed form is to do so on the form itself and have the amendment signed or initialled by the parties. Alternatively, suitably amended or special clauses can be annexed to the form, duly signed by the parties, or the second part of clause 1.3 itself can be deleted. Were it not for this clause, the general law would give effect to any amendments which were contained in the bills, etc. This clause is therefore of great significance. Consideration should be given to the possibility of deleting clause 1.3 and allowing the sensible interpretation offered by the general law to take effect.

3.1.3 Errors

Clauses 2.12, 2.13 and 2.14 further stress the importance of having a thoroughly prepared set of contract documents and of using the same care to include anything which may be issued to the contractor to assist it in carrying out the works.

The architect is obliged to issue instructions (see section 4.1.4) to correct any of the following:

- Departures from the method of preparation of the bills of quantities referred to in clause 2.12.1: this states that the bills of

quantities must be prepared in accordance with the Standard Method of Measurement, except where certain items are specifically stated to have been measured in a different way. This is always a fruitful source of claims by the contractor
- Errors in description or quantity or omissions of any item in any of the contract documents, including items which are the subject of a provisional sum for defined work. If the information for a provisional sum for defined work is missing, it must be provided
- Inconsistencies which occur within any one of the contract documents or between the contract documents: this obligation extends to inconsistencies which may occur following the issue of architect's instructions (except, obviously, a variation instruction), further information under clauses 2.10 and 2.11 or levels and setting-out information under clause 2.9. The contractor is obliged, under clause 2.13.3, to give written notice to the architect on finding any such inconsistency. However, that does not mean that the contractor is obliged to look for them: *London Borough of Merton* v. *Stanley Hugh Leach Ltd* (1985).

None of these errors, inconsistencies, or departures will invalidate the contract, but if an instruction changes the quality or quantity of works as understood from clause 4.1, or in any other way constitutes a variation, it must be valued under section 5.

There is a crumb of comfort to be gained from the fact that it is established law that, if something is omitted from the bills or specification which it is quite clear to everybody should be there and is necessary, the contractor will be deemed to have included it in its price: *Williams* v. *Fitzmaurice* (1858). There is scope for dispute in applying this principle. The contractor will always maintain that it certainly was not clear, or that it assumed that the employer was going to employ someone else to do or supply the thing in question. In the context of a particular contract, however, it should be possible to make a fair decision on any particular item. A clause to the effect that everything necessary to complete the work should be included by the contractor will tend to settle the matter in most cases.

An example will clarify the point. Assume that the bills of quantities provide for the contractor to supply damp-proof course of a particular quality and in a given quantity, but the item requiring it to be laid has been inadvertently omitted. The contractor will be deemed to have included the laying in its price because it is clear to everyone that laying is required. On the other hand, if the architect fails to include the supply and fixing of, say, door furniture, the contractor may have a point in claiming that it assumed that the employer wanted to carry out this operation. Moreover, the contractor will have had no information at all on which to base any price, and it would be entitled to its extra costs. In practice, the extreme situation should never arise, because the contractor would certainly query the omission of door furniture at tender stage.

3.1.4 Custody and copies

Clause 2.8.1 makes it quite clear that the contract documents must remain in the custody of the employer. There is a proviso that the contractor must be allowed to inspect them at all reasonable times, but this provision appears to be redundant in the light of the fact that under clause 2.8.2 the architect must provide the contractor with a copy of the documents certified on behalf of the employer. It is essential that the copy is the same as the original. Very often, two sets of documents are prepared and all are signed by both parties, but contrary to the belief of some contractors this is not strictly necessary. The employer should check that the copy is the same as the original and sign the certificate, because that is the employer's responsibility. The certificate is usually inscribed on each document, including the drawings. Some architects favour a very elaborate pseudo-legal turn of phrase, but it is sufficient to state, 'I certify that this is a true copy of the contract document.'

The clause imposes a further duty on the architect to supply the contractor with two copies (uncertified) of each of the contract documents.

The further drawings and details which are to be supplied under clauses 2.10 and 2.11 are not contract documents. They are intended merely to amplify the information contained in the contract documents. The drawings must be accurate, and any revisions must be clearly identified: *London Borough of Merton* v. *Stanley Hugh Leach Ltd* (1985). The obligation of the architect is to supply only such drawings and details as are reasonably necessary to enable the contractor to carry out and complete the Works in accordance with the conditions: i.e. among other things, to complete by the stipulated completion date or any extended date. It was thought that the architect was not entitled to delay provision of the details simply because it appeared unlikely that the contractor would be ready for them before the date for completion. That is not the current position (see section 5.2.2). However, note that the contractor has grounds for extension of time or loss and/or expense if the architect delays the contractor (see section 10.2.4).

3.1.5 Limits to use

Clause 2.8.3 contains safeguards for both contractor and architect. It prohibits the use for any purpose other than the contract of any documents (contract documents or others) issued in connection with the contract. It also prohibits the employer, the architect, and the quantity surveyor from using any of the contractor's rates or prices in the contract documents, the contract sum analysis, or the schedule of rates for any purpose other than the contract. None of the contractor's rates or prices must be divulged to third parties.

The prohibition against the use of documents by the contractor simply states expressly what is understood from the general law with regard to the architect's copyright.

The prohibition against divulging the contractor's rates is designed to protect the contractor's most precious possession – the ability to tender competitively and thus secure work. To divulge the rates to a competitor is probably one of the most harmful things which could be done to any contractor. In practice, it is extremely difficult for the contractor to ensure that such prices are not used, for example by the quantity surveyor to help estimate some other current job.

There is no requirement for the contractor to return any drawings or details at the end of the contract. There is nothing to prevent the architect from asking for them in order to be sure that they will not be used for any other purpose, but the architect cannot require the contractor to return the certified copy of the contract documents, which will be needed for the contractor's own records.

3.1.6 Notices

Clause 1.17 of the contract sets out the requirements for the giving or service of notices or other documents. It only applies if the contract does not expressly state the way in which service of documents is to be achieved. Therefore it does not apply to notices given in connection with the termination procedures in section 8, because that clause states that service is to be carried out by means of actual delivery, special or recorded delivery. In other cases, service is to be by any effective means to the address given in the contract particulars or to any agreed address. If the parties cannot agree over service and appropriate addresses, service can be achieved by addressing the document to the last known principal business address or, if the addressee is a body corporate, to that body's registered office or its principal office, provided it is prepaid and sent by post.

Clause 1.5 sets out the way in which periods of days are to be reckoned in order to comply with the Housing Grants, Construction and Regeneration Act 1996. If something must be done within a certain number of days from a particular date, the period begins on the day after that date. Days which are public holidays are excluded. Public holidays are defined in clause 1.1 as 'Christmas Day, Good Friday or a day which under the Banking and Financial Dealings Act 1971 is a bank holiday'. A footnote instructs the user to amend the definition if different public holidays apply.

The law applicable to the contract is to be the law of England, no matter that the nationality, residence or domicile of any of the parties is elsewhere (clause 1.12). Where a different system of law is required, this clause must be amended. Therefore, if the Works are being carried out in Northern Ireland, the parties will wish the applicable law to be the law of Northern Ireland. Curiously, the applicable law of the two bonds in schedule 3 of the contract has been stated to be the law of England and Wales since they were first introduced, and there is no specific note to amend. However, parties who amend the law of the contract will doubtless, for consistency, wish to amend the applicable law of the bonds also. Although it may be argued that there is currently no real difference between the two jurisdictions, that may well change.

Electronic communication may be employed if the parties so wish. They can insert an apropriate reference into the contract particulars and then clause 1.8 states that those procedures apply. The intention is that the particular communications which may be sent electronically should be listed. For example, the parties may decide that all correspondence may be by e-mail or even that architect's instructions and some certificates may be sent this way. However, traditional hard copy should be used for important documents such as the final certificate and notice of adjudication or arbitration. In practice, it is also sensible to ensure that all certificates are issued in hard copy. If issued in both electronic and hard format, there must be no doubt about the date of issue, which has important contractual implications.

Copies of all electronic communications must be kept on file.

3.2 *Insurance*

3.2.1 Indemnity

The indemnity and insurance provisions were completely revised by amendment 1 to IFC 84 in November 1986 and again to a more limited extent by amendment 10 in July 1996. This new contract deals with insurance under clause 6 with Works insurance options in schedule 1, near the back of the form. The provisions do not seem to have been significantly changed. Two new definitions have been introduced into clause 1.1: 'Contractor's Persons' and 'Employer's Persons'. These replace the former rather long-winded descriptions referring to servants, agents and persons for whom the contractor or the employer is responsible.

Under clauses 6.1 and 6.2 the contractor assumes liability for, and indemnifies the employer against, any liability arising out of the carrying out of the works in respect of the following:

- Personal injury or death of any person, unless and to the extent due to act or neglect of the employer or of the employer's persons. The persons will include anyone employed and paid by the employer, such as directly employed contractors, the clerk of works, and the architect
- Injury or damage to any kind of property arising from the Works, but other than the Works themselves and materials on site intended for incorporation, provided and to the extent that it is due to the negligence, breach of statutory duty, omission, or default of the contractor or of any of the contractor's persons; the contractor's liability is limited, as compared with its liability for personal injury or death, because the contractor must be at fault for the indemnity to be operative. The contractor has no liability under this clause for any loss or damage which the employer is to insure under paragraph C.1 of schedule 1 (existing buildings). Therefore, where work is being carried out to existing premises, the contractor will not be liable for damage to the existing premises or their contents, because they will already be insured under paragraph C.1 of schedule 1.

Clause 6.3 deals expressly with the exclusion of the Works and site materials from these indemnities. The employer is responsible for the injury or death of any person only insofar as the injury or death was caused by the employer's or the employer's person's act or neglect. The employer is responsible for all loss or damage to property except to the extent that it is caused by the contractor's or contractor's person's default. In practice, if any claim does arise, it is likely to be the employer who will be sued. In turn, the employer will join the contractor as a third party in any action and claim an indemnity under this clause.

3.2.2 Injury to persons and property

Clause 6.4 requires the contractor to take out insurance (unless its existing insurances are adequate) to cover its liabilities under clauses 6.1 and 6.2. This requirement is stated to be without prejudice to its liability to indemnify the employer under these clauses. Thus the fact that the contractor has taken out or maintains the appropriate insurance cover does not affect its liabilities. If, for some reason, the insurance company refused to pay in the case of an incident, the contractor would be obliged to find the money itself.

The insurance cover must be for a sum not less than whatever is stated in the contract particulars to the contract for any one occurrence or series of occurrences arising out of one event. The insurance against claims for personal injury or death of an employee or apprentice of the contractor must comply with all relevant legislation.

The employer has the right (clause 6.4.2) to inspect documentary evidence that the contractor is maintaining proper insurance cover, and in particular to inspect policies and premium receipts. The clause requires these to be sent to the architect for inspection by the employer. The employer is not to exercise this right unreasonably or vexatiously; it is likely to be enforced at the beginning of the contract and at the time of any required premium renewal.

It is always wise for the employer to retain the service of an insurance broker in connection with all the insurance provisions of the contract. It is for the employer to appoint such a person on the architect's advice. The broker should inspect the relevant documents and confirm in writing that they comply with the contract requirements. On no account should the architect simply send the policy to the employer without comment: *Pozzolanic Lytag* v. *Brian Hobson Associates* (1999). It seems that the architect has three options: to give advice to the employer, to obtain advice and pass it on, or to advise the employer to seek specialist insurance advice. As noted above, the last is probably the most sensible option.

If the contractor fails to insure under clause 6.4.1, the employer has the right (clause 6.4.3) to take out the appropriate insurance and to deduct the amount of any premium from monies due or to become due to the contractor. (Note that in the case of deduction the relevant notice under clauses 4.8.2 and 4.8.3 should be given. These are the notices relevant to interim certificates. In view of the importance of the insurance it seems unlikely, to say the least, that the employer will need to use the notices relevant to the final certificate. If that is the case, there is something seriously wrong with the contract administration.) Alternatively, the employer may recover the amount from the contractor as a debt. It is essential that the employer exercises that right, and such is the importance of maintaining continuous insurance cover, the architect should lose no time in advising the employer if it is discovered that the contractor has defaulted. The cover should be effected through and on the advice of the employer's broker.

The architect must immediately write to the contractor and the employer. The exact circumstances may vary widely, but the letters in Figures 3.1 and 3.2 are examples.

3.2.3 Things which are the liability of the employer

Clause 6.5.1 provides for insurance against damage caused by the carrying out of the Works when there is no negligence or default by any party. The clause is operative only if it is stated in the contract particulars that the insurance may be required. The contractor, if so instructed by the architect, must maintain the insurance in the joint names of the employer and the contractor. The amount of cover must be specified in the contract particulars, and a footnote advises that it should not expire until the end of the rectification period. Realistically, because the contractor will be rectifying work after that date, the insurance ought to extend until the contractor completes making good of defects. Liability envisaged by this clause covers damage to any property, other than the Works and site materials, caused by collapse, subsidence, heave, vibration, weakening or removal of support, or lowering of groundwater resulting from the Works. Not covered is damage:

- For which the contractor is liable under clause 6.2
- Due to errors or omissions in the architect's design
- Reasonably foreseen to be inevitable
- Which it is the responsibility of the employer to insure under paragraph C.1 of schedule 1
- Due to nuclear risk, or sonic booms
- Arising from war, hostilities, civil war etc.
- Caused directly or indirectly by pollution unless occurring suddenly as a result of an unexpected incident
- To the Works and site materials unless they or parts of them are the subject of a certificate of practical completion
- Which results in the employer paying damages for breach of contract.

Figure 3.1
Architect to contractor if contractor fails to maintain clause 6.4.1 insurance cover

Dear Sir

I refer to my telephone conversation with your [*insert name*] this morning and confirm that you are unable to produce the insurance policy, premium receipts or documentary evidence that the insurances required by clause 6.4.1 are being maintained.

In view of the importance of the insurance and without prejudice to your liabilities under clauses 6.1 and 6.2 of the conditions of contract, the employer is arranging to exercise appropriate rights under clause 6.4.3 immediately. Any sum or sums payable by the employer in respect of the premiums will be deducted from any monies due or to become due to you or will be recovered from you as a debt.

Yours faithfully

Copy: Employer
Quantity surveyor

Figure 3.2
Architect to employer if contractor fails to maintain clause 6.4.1 insurance cover

Dear

The contractor is unable to provide evidence to show that it is maintaining the insurances required by clause 6.4.1 of the conditions of contract.

In view of the importance of the insurance, you should immediately instruct your broker to provide the necessary cover, effective immediately. You are entitled to do this under clause 6.4.3. You are also entitled to deduct the amount of the premium from your next payment to the contractor or, alternatively, you may wish to recover it as a debt. If you intend to deduct, remember that you must serve the appropriate notices under clauses 4.8.2 and 4.8.3. [*Add if appropriate:*] In this instance simple deduction would appear to be the easiest method of recovery.

A copy of my letter to the contractor, dated [*insert date*], is enclosed for your information.

Yours faithfully

The contractor must obtain the employer's approval of the insurers proposed, but not, apparently, to the amount of the premium. One method of sorting this out is for the employer to place the matter in the hands of the broker. That advice will be needed in any event. If the employer later indicates to the contractor that approval will be forthcoming for the insurer recommended by the broker, the contractor will probably be delighted to place the insurance with whomsoever the employer wishes. The device of the 'double letter' (Figures 3.3 and 3.4) is useful in such circumstances, as it leaves the responsibility where it belongs – with the contractor. The two letters are sent at the same time.

The contractor must deposit the policy and premium receipts with the employer. The amounts paid by the contractor in premiums under this clause are to be added to the contract sum. If the contractor fails to insure or to maintain the insurance under this clause, the employer may take out the insurance, in which case nothing is added to the contract sum under this clause. Since the employer, or certainly the architect, knows what insurance is needed and with which insurer it is desired to place the cover, there may be something to be said for amending this clause to make the employer responsible for insuring under this clause.

There is a general proviso, contained in clause 6.6, stating that the contractor shall not be liable to indemnify the employer or to insure against damage to the Works, the site, or any property due to nuclear perils and the like. This clause overrides anything contained in clauses 6.1 to 6.4. These are referred to as 'Excepted Risks', and a full definition is to be found in clause 6.8.

There are two very important definitions in clause 6.8 and 1.1: 'all risks insurance' and 'site materials'. The former states the risks for which insurance is required, and the latter refers to all unfixed materials and goods delivered to, placed on or adjacent to the works and intended for incorporation therein. The key point about 'all risks insurance' is that it must provide cover against 'any physical loss or damage to work executed and Site Materials'. Cover is not limited to 'specified perils' (also defined in clause 6.8). It includes other risks such as impact, subsidence, theft and vandalism. The distinction between the two types of risk has to be carefully noted because the contract very often limits the risks to 'specified perils' (for example when an extension of time is being considered).

'Specified perils' are fire, lightning, explosion, storm, tempest, flood, escape of water from any water tank, apparatus or pipes, earthquake, aircraft and other aerial devices or articles dropped therefrom, riot and civil commotion, excluding any loss or damage

Figure 3.3
Architect to contractor regarding clause 6.5.1 insurance cover (double letter 1)

Dear

I should be pleased if you would inform me of the name of the insurers with whom you intend to place insurance in accordance with clause 6.5.1 of the conditions of contract. The contract provides that the employer must give approval before you proceed to place the insurance.

Yours faithfully

Copy: Employer

Figure 3.4
Architect to contractor regarding clause 6.5.1 insurance cover (double letter 2)

WITHOUT PREJUDICE

Dear

I write with regard to the insurer whose name you are required to submit to the employer under clause 6.5.1 of the conditions of contract. I am instructed by the employer that, if you were to suggest [*insert the name of the insurers recommended by the employer's broker*], the employer would be prepared to approve them.

Yours faithfully

Copy: Employer

caused by ionising radiations or contamination by radioactivity from any nuclear fuel or from any nuclear waste from the combustion of nuclear fuel, radioactive toxic explosive or other hazardous properties of any explosive nuclear assembly or nuclear component thereof, pressure waves caused by aircraft or other aerial devices travelling at sonic or supersonic speeds.

3.2.4 Insurance of the Works: alternative clauses

Clause 6.7 provides that alternative Works insurance clauses to deal with loss and damage to the Works and unfixed materials and goods are set out in schedule 1. The applicable insurance option is to be stated in the contract particulars.

Schedule 1 refers to alternative clauses, which are:

- Option A – As far as new buildings are concerned, this is the most usual provision. The contractor is required to take out all risks insurance
- Option B – This deals with the situation if the employer wishes to take out insurance for new building work
- Option C – This deals with the insurance of Works in or extensions to existing buildings. The employer is required to insure the existing structure and contents and the new work.

One or other of these options must apply, as stated in the contract particulars.

Schedule 1 is divided into three parts, two of which do not apply. So, if the contractor is to insure a new building, options B and C do not apply. If, rarely, the employer is to insure a new building, options A and C do not apply. Options A and B do not apply if the employer is to insure existing structures and new work to them.

3.2.5 A new building where the contractor is required to insure

Option A covers this situation. Paragraph A.1 obliges the contractor to take out and maintain a joint names policy for all risks insurance. The list of risks is extensive, and a footnote advises that some of the risks may not be able to be covered. The matter should be agreed at tender stage and appropriate amendments made to the clause. The insurance must cover:

- The full reinstatement value of the Works; and
- The cost of any professional fees, expressed as a percentage. The percentage must be inserted in the contract particulars by the employer. Failure to make the insertion would probably result in the cost of professional fees being borne by the employer.

Care must be taken in regard to 'full reinstatement value'. It should be remembered that the Works will increase in value as the contract progresses. The sum insured must reflect the actual cost of reinstatement of the Works and any lost or damaged site materials, which will include the cost of removing debris. Note that all risks insurance does not cover what is referred to as 'consequential loss': *Kruger Tissue (Industrial) Ltd* v. *Frank Galliers Ltd and DMC Industrial Roofing & Cladding Services (A Firm) and H.&H. Construction (A Firm)* (1998); *Horbury Building Systems Ltd* v. *Hampden Insurance N.V.* (2004). An example of this would be where the employer suffered loss because of the increased costs of carrying out work not completed at the time of the damage. The contractor can only recover the actual insured sum. If it proves to be insufficient, it has to bear the excess. The policy must be maintained up to and including practical completion of the Works, or the date of termination of the contractor's employment under section 8 (whether or not the validity is contested), whichever is the earlier.

The joint names policy is to be taken out with insurers approved by the employer. The policy, premium receipts and any relevant endorsements must be sent to the architect, who must deposit them with the employer. If the contractor defaults in taking out or in maintaining the insurance, the employer is entitled to take out the insurance. In that situation, the employer may deduct the amounts of any premiums from the sums otherwise due to the contractor (e.g. certified amounts), or they may be recovered as a debt. It is unlikely that any employer will take the latter course when the former is so readily available.

Contractors usually have general insurance to cover the risk of damage to the Works. This type of insurance is acceptable under the terms of the contract if the policy provides 'all risks' cover for no less than full reinstatement value and the appropriate percentage of professional fees, and it is in joint names, and the contractor can send documentary evidence that the insurance is in force. If the contractor defaults, the employer may take out and maintain a policy and deduct the cost as before.

If loss or damage occurs due to one of the risks covered by the policy, the contractor is obliged to notify the architect and the employer under the provisions of paragraph A.4. The contractor must do this as soon as the loss or damage is discovered.

Paragraph A.4.2 stipulates that the occurrence of loss or damage is to be disregarded in computing amounts payable to the contractor under the contract. This paragraph covers the position where the architect issues a certificate which includes work which is later damaged. Later certificates are not to allow for the damaged work. Such work must be treated, for certification purposes, as if it was undamaged. Thus the architect may have to certify, as work properly done, work which has been destroyed by one of the insured risks.

There is nothing strange in this. Without this provision, the contractor must restore damaged work after any inspection which the insurer may require has taken place. The effect of this is that the contractor could be under a duty to restore the work before the insurance monies have been paid. Paragraph A.4.4 states that the contractor and the sub-contractors must authorise the insurers to pay the insurance monies to the employer, who must pay all the money received, less only professional fees, to the contractor in instalments under the architect's certificates. It is good to see that the new form has adopted a better formula of words to define the amount than the complex clause formerly used.

3.2.6 A new building where the employer insures

This is not very common, but where the employer wishes to insure, option B deals with the situation. It is in very similar terms to option A. The employer must take out insurance in joint names for the full reinstatement value of the Works together with an appropriate percentage for professional fees. The contractor must give notice as before if there is any loss or damage, but a very important difference from the position under option A is that the restoration and repair work is to be treated as a variation. The result of this is that, although the employer is entitled to be paid the full amount of any insurance money, the employer must stand any further expense. The contractor, on the other hand, receives the full cost of repair.

3.2.7 Alterations or extensions to an existing building

Option C deals separately with loss or damage to existing structures and contents and with loss or damage to the Works carried out to the structures or as extensions. If an existing building is involved, there is no option. The employer must take out and maintain insurance.

Existing structures and contents owned by the employer, or for which the employer is responsible, must be insured under paragraph C.1. The risks to be covered are specified perils. If part of an extension is taken into the employer's possession, the relevant part will form part of the existing structures for insurance purposes from the relevant date. Loss or damage to the Works is covered by paragraph C.2. It is similar to the position under option B against 'all risks'. Both sets of insurance must be in joint names, and the contractor authorises payment of insurance money directly to the employer. In common with option B, if the insurance payment is insufficient to cover the cost of restoration, the employer must stand the amount of shortfall. It has been held that fire caused by the contractor's negligence is not covered by this insurance, but must be covered by the employer's own insurance: *London Borough of Barking and Dagenham* v. *Stamford Asphalt Company* (1997). Although concerned with the Minor Works Form of Contract (MW 80), the principle should be capable of application to IC. However, it does appear to cut across generally accepted principles that the insurance is to deal with damage however caused, and calls into question the purpose of insurance in joint names.

Except where the employer is a local authority, the contractor is entitled to demand proof that the insurances have been taken out and maintained. In default, the contractor may take out the appropriate insurance itself and the amount it pays is to be added to the contract sum (paragraph C.3.1.3). The position is the same under paragraph B.2.1. The contractor has an important extra power under paragraph C.3.1.2. It has right of entry as may be required to make a survey and inventory of the existing structures and contents.

The contractor must notify the employer and the architect in writing upon discovering loss or damage covered only under the joint names policy taken out under paragraph C.2. If neither party then terminates the contractor's employment (see section 12.1.6), the contractor must proceed to carry out appropriate restoration work after the insurers have carried out any inspection they may require.

There is no provision for notifying the employer and the architect if loss or damage to existing structures is discovered. It is treated as a matter for the employer and outside the contract procedures.

3.2.8 Benefits for sub-contractors

The joint names policies required under paragraphs A.1, A.3, B.1 or C.2 of schedule 1 must either provide for recognition of each sub-contractor as an insured or include a waiver of the insurer's rights of subrogation against any sub-contractor in respect of loss or damage by any of the specified perils (clause 6.9.1). This is a very useful provision for sub-contractors who, if responsible for the loss, may otherwise face a writ from the insurers. The provisions are extended to embrace joint names policies under paragraphs A.2, B.2.1.2 or C.3.1.2 of schedule 1. If paragraph C.1 applies, the employer must ensure that a named sub-contractor is recognised as an insured or has the benefit of waiver of the insurer's right of subrogation.

3.2.9 Joint fire code

In the definitions the 'Joint Fire Code' is said to be the 'Joint Code of Practice on the Protection from Fire of Construction Sites and Buildings Undergoing Renovation published by the Construction Confederation and the Fire Protection Association with the support of the Association of British Insurers, the Chief and Assistant Fire Officers Association and the London Fire Brigade as amended from time to time'. The code states that non-compliance could result in insurance ceasing to be available.

If the insurer of the Works requires compliance with the code, the contract particulars entry should so state. Special requirements apply if the insurer categorises the Works as a 'Large Project', and the contract particulars must be completed appropriately.

Clause 6.12 requires both employer and contractor and anyone employed by them or anyone on the Works to comply with the code.

Clause 6.13 makes clear that if there is a breach of the code and the insurer requires, by notice, remedial measures, the contractor must ensure that the measures are carried out by the date specified in the notice and in accordance with the architect's instructions, if any. If the contractor does not begin the remedial measures in seven days from receipt of the notice, or fails to proceed regularly and diligently, the employer may employ and pay others to do the work. In that case an appropriate deduction is to be made from the contract sum.

If the code is amended after the base date and the contractor is thereby put to additional cost, such cost must be added to the contract sum.

3.2.10 Terrorism cover

If the insurers notify either party that terrorism cover will cease from a specified date, either the employer or the contractor, as appropriate, must notify the other. The employer has two options: either to require that the Works continue to be carried out or to specify that the contractor's employment will terminate on a date after the date of the insurer's notification, but before the date of cessation. The employer's notice must be in writing. If termination is the option, clauses 8.12.2 to 8.12.5 apply (excluding clause 8.12.3.5). If the decision is not to terminate, and damage is suffered as a result of terrorism, the resulting work is treated as a variation.

3.3 *Summary*

Contract documents

- Contract documents are the only evidence of the contract
- They may consist of whatever the parties agree
- Drawings prevail over the specification
- Any quantities prevail over the drawings
- Printed conditions prevail over all
- Errors, inconsistencies and departures are to be corrected by architect's instructions and a variation allowed if appropriate
- Items missing from the specification may be deemed to be included if it is obvious to all that they should be there
- Contract documents must be kept by the employer, with a true copy to the contractor
- No other drawing or detail is a contract document
- Contract documents and any other documents issued for the contract must not be used for any other purpose
- Contractor's rates must not be divulged or used for another purpose.

Insurance

- Employer's indemnity covers personal injury and death and damage to property, subject to certain exceptions
- Contractor must insure to cover the indemnities
- Employer has the right to insure if contractor fails
- Special insurance can be taken out to cover instances where there is no default by any party
- Either party may insure new work against all risks
- Only the employer may insure existing structures against specified perils and new work thereto against all risks
- Where terrorism cover is ended, the employer may proceed or terminate.

CHAPTER FOUR
THE ARCHITECT'S AUTHORITY AND DUTIES

4.1 *Authority*

4.1.1 General

The extent of the architect's authority depends on the agreement with the employer. The wise architect will have entered into a formal written contract, preferably incorporating the terms of the RIBA Standard Form of Agreement for the Appointment of an Architect (SFA 99), or the Conditions of Engagement (CE/99), which are simpler. The architect's powers and duties under IC (see Table 4.1) flow directly from the agreement with the employer, not from the building contract itself, to which the architect is not a party. It follows that if the architect fails to carry out the relevant duties properly, the employer, but not the contractor, can take legal action against the architect under the terms of engagement with the employer. It is possible for the contractor to bring an action directly against the architect in tort for negligence, but it must be shown that the architect owed a legal duty of care to the contractor, that the architect was in breach of that duty, and that, by reason of the breach, the contractor suffered loss or damage. It was thought that the chances of the contractor being successful in such a contention had been severely reduced, if not totally extinguished, following *Pacific Associates Inc* v. *Baxter* (1988), where an engineer was found to have no duty of care to the contractor when administering the contract. The engineer's duty was to the employer, from whom the contractor could seek redress; but see below.

Generally, redress requires the contractor to take action against the employer under the contract, the employer in turn taking action against the architect. If the architect fails to carry out any duties under the contract, this is generally a default for which the employer may be held responsible: *Croudace Ltd* v. *London Borough of Lambeth* (1986). However, it seems that the employer's responsibility is not triggered

Table 4.1
Architect's powers and duties under IC and ICD (including powers and duties in the schedules)

Clause	Power/duty	Precondition/comment
1.9	**Duty** Issue certificates to employer with copy to contractor	Unless provided otherwise
2.8.2	**Duty** Provide contractor with copies of contract documents and drawings, etc.	Immediately after execution of contract
2.8.3	**Duty** Not to divulge or use rates in priced document	
2.9	**Duty** Determine levels and provide contractor with accurately dimensioned drawings, etc.	To enable contractor to set out the Works
2.10	**Duty** Ensure that two copies of the information listed in the information release schedule are released in due time	Employer and contractor may agree to vary the release times
2.11	**Duty** Provide contractor with two copies of additional information	Without charge, if not included in the information release schedule
2.13.1	**Duty** Issue instructions	Regarding departure, error or omission in contract bills and any error in description or quantity, any omission or any inconsistency in or between contract documents, architect's instructions and drawings
2.15.1	**Duty** Give a written notice to the contractor specifying the divergence	If the architect finds a divergence between statutory requirements and any contract documents or instructions
2.15.2	**Duty** Give instructions regarding divergence	Within 7 days of finding a divergence (or under ICD within 14 days of the contractor's proposed amendment to the CDP documents). If the instruction requires the Works to be varied, it must be treated as a variation

Table 4.1 *Cont'd*

Clause	Power/duty	Precondition/comment
2.17	**Power** Consent in writing to the removal of unfixed goods or materials	The consent must not be unreasonably withheld or delayed
2.19.1	**Duty** Make in writing a fair and reasonable extension of time	As soon as the architect is able. If the contractor has given written notice in due time and the architect has formed the opinion that the Works or section is being or likely to be delayed beyond the relevant completion date
2.19.2	**Duty** Make in writing a fair and reasonable extension of time	As soon as the architect is able. If certain Relevant Events occur after the relevant completion date but before practical completion
2.19.3	**Power** Make in writing a fair and reasonable extension of time	Up to 12 weeks after practical completion of the Works or section whether upon reviewing a previous decision or otherwise and whether or not the contractor has given notice
2.21	**Duty** Forthwith issue a certificate of practical completion of the Works or a section	When practical completion has been achieved and the contractor has sufficiently complied with clause 3.18.3
2.22	**Duty** Issue a certificate of non-completion	If the contractor fails to complete the Works or section by the relevant completion date
2.25	**Duty** Issue a written statement to the contractor on behalf of the employer identifying the part(s) taken into possession and giving the date(s)	If the employer takes possession of a part of the Works or section before practical completion with the consent of the contractor
2.27	**Duty** Issue a certificate of making good	When the architect is of the opinion that defects in the part, required to be made good, have been made good

Table 4.1 *Cont'd*

Clause	Power/duty	Precondition/comment
2.30	**Duty** Notify the contractor of defects in a schedule of defects. **Power** Instruct that certain defects are not to be made good	Not later than 14 days after the expiry of the rectification period. If the employer consents. An appropriate deduction is to be made from the contract sum
2.31	**Duty** Issue a certificate of making good	When the architect is of the opinion that defects, required to be made good, have been made good
3.1	**Power** Access to the Works and work being prepared	At all reasonable times and subject to reasonable restrictions to protect proprietary rights
3.5	**Power** Consent to sub-letting any part of the Works (or under ICD the design)	Consent must not be unreasonably delayed or withheld, but the contractor remains wholly responsible
3.8	**Power** Issue instructions empowered by the contract	The contractor must comply forthwith unless it objects to a clause 5.1.2 variation (under ICD it may object if the contractor is of the opinion that the instruction injuriously affects the efficiency of the design of the CDP)
3.9	**Power** Issue written notice to the contractor requiring compliance with an instruction	The contractor has 7 days to comply or the employer may engage others to do the work and an appropriate deduction will be made from the contract sum
3.10	**Duty** Forthwith comply with a request from the contractor to specify the empowering clause	If contractor then complies, the instruction is deemed duly given under that clause
3.11	**Power** Issue instructions requiring a variation Sanction a variation by the contractor	No architect-authorised variation will vitiate the contract. Clause 5.1.2 instructions subject to contractor's right of objection. Under ICD, a variation instruction is an alteration to the Employer's Requirements

Table 4.1 *Cont'd*

Clause	Power/duty	Precondition/comment
3.12	**Power** Issue instructions requiring postponement of work	
3.13	**Duty** Issue instructions regarding provisional sums	
3.14	**Power** Issue instructions requiring opening up or testing	Cost to be added to contract sum unless the work is not in accordance with the contract
3.15.1	**Power** Issue instructions following failure of work	If architect *either* has not received contractor's proposals within 7 days of discovery of failure, *or* is not satisfied with contractor's proposed actions, *or* safety considerations or statutory obligations require urgent action
3.15.2	**Power** Withdraw or modify instruction under clause 3.15.1	If contractor objects to compliance within 10 days of receipt of 3.15.1 instruction and architect accepts contractor's reasons
3.16.1	**Power** Issue instructions requiring the removal of work not in accordance with the contract	
3.16.2	**Power** Issue instructions as are reasonably necessary in consequence	If work not carried out in a proper and workmanlike manner
3.17	**Power** Instruct the exclusion of persons from the site	Not unreasonably or vexatiously
4.6.1	**Duty** Certify the amount of interim payments	At the dates in the contract particulars. Subject to agreement on stage payment
4.9	**Duty** Certify payment of the percentage stated in the contract particulars of the total value of the Works and 100% of clause 4.7.2 amounts less relevant deductions	Within 14 days after practical completion

Table 4.1 *Cont'd*

Clause	Power/duty	Precondition/comment
4.14.1	**Duty** Issue a final certificate	Within 28 days of the latest of: • sending computations of adjusted contract sum to contractor • issuing certificate of making good
4.17	**Duty** Ascertain or instruct the quantity surveyor to ascertain amount of loss and/or expense incurred	If contractor makes written application within reasonable time of the incurring becoming apparent; if the contract does not otherwise provide; if the contractor has provided in support such information to the architect and/or the quantity surveyor as reasonably necessary
5.4	**Duty** Verify vouchers for daywork	If delivered by the contractor not later than the end of the week following the week in which the work was carried out
6.5.1	**Power** Instruct the contractor to take out policy of insurance indemnifying the employer against claims, etc. in respect of collapse, subsidence, heave, vibration, weakening or removal of support or lowering of ground water as a result of the Works	If the contract particulars so state
6.15.3 (ICD only)	**Power** Reasonably request the contractor to produce documentary evidence that CDP professional indemnity insurance has been affected	

Table 4.1 *Cont'd*

Clause	Power/duty	Precondition/comment
8.4.1	**Power** Serve written notice on the contractor by special, recorded or actual delivery	If the contractor without reasonable cause wholly or substantially suspends the carrying out of the Works before completion; *or* fails to proceed regularly and diligently with the Works; *or* refuses or neglects to comply with a written notice from the architect requiring the removal of defective work or materials or goods and by such refusal or neglect the Works are materially affected; *or* fails to comply with clauses 3.5 (sub-letting), 3.7 (named sub-contractors) or 7.1 (assignment); *or* fails to comply with the CDM Regulations
8.7.3	**Duty** Issue certificate showing expenses incurred by employer, amount of payments to the contractor and the total amount that would have been payable to the contractor	Within a reasonable time after completion of the Works and defects following determination by the employer. Alternatively, the employer may set out the information in the form of a statement
Schedule 2 para 2	**Power** Instruct • to change particulars to remove impediment • to omit work • to omit work and substitute a provisional sum	In regard to named sub-contractors
Schedule 2 para 4	**Power** Instruct that work is to be carried out by person other than named in the contract documents	The instruction must omit work and substitute a provisional sum
Schedule 2 para 5.1	**Power** Require work to be executed by person named in an instruction	The person must then be employed by the contractor as a named sub-contractor

Table 4.1 *Cont'd*

Clause	Power/duty	Precondition/comment
Schedule 2 para 6	**Power** Consent to the contractor accepting termination or repudiation of the sub-contract by the named sub-contractor	
Schedule 2 para 7	**Power** Give instructions: • to name another sub-contractor • to require the contractor to make its own arrangements • to omit the work	As necessary after termination of employment of named sub-contractor
Schedule 2 para 8	**Power** Consent to termination of the named sub-contractor's employment under clause 7.4, 7.5 or 7.6 of ICSub/NAM/C	

until the employer becomes aware that the architect is not performing properly: *Penwith District Council* v. *VP Developments Ltd* (1999).

Traditionally, the architect's role changes when the contract between employer and contractor is signed. Up to that time, the architect has been acting as agent, in a limited capacity, for the employer. After the signing, the architect assumes a dual and difficult role: still an agent of the employer, but also charged with seeing that the terms of the contract are administered fairly (*London Borough of Merton* v. *Stanley Hugh Leach Ltd* (1985)); 'without fear or favour' is the usual term.

'Without favour' is absolutely right, but 'without fear' is obsolete nowadays. It used to be thought that, when architects carried out their duties under the contract, they were immune from any action against them in negligence by either party. If such a state ever really existed, it was changed in 1974 by the celebrated case of *Sutcliffe* v. *Thackrah* (1974). There is now no doubt that architects are open to an action for negligence on every decision they take; certainly from the employer's point of view they are not acting in a 'quasi-arbitral' capacity, notwithstanding that they are still required to act fairly between the parties in administrating the contract. The situation is totally unrealistic, because there may be instances when architects have to decide whether they themselves are in default, and act accordingly – for example, in cases of claims for extension of time or for loss and/or expense. If they issue instructions late, they are duty bound – if contractors make proper claims under the contract – to award an extension or ascertain a financial claim, as the case may be. Employers may well be able to recover that loss from architects. Architects' tortious liabilities may exceed those contained in the conditions of engagement: *Holt* v. *Payne Skillington* (1995). So far as actions by the contractor against the architect are concerned, the courts seem prepared to extend the scope of *Hedley Byrne & Partners* v. *Heller & Co Ltd* (1963), liability for negligent misstatements, to include negligent actions: *Henderson* v. *Merrett Syndicates* (1994); *Conway* v. *Crowe Kelsey and Partners* (1994); *Cliffe Holdings* v. *Parkman Buck Ltd and Wildrington* (1996). The principles of reliance and voluntary assumption of liability to another party are being emphasised. A contractor was partially successful in suing the architect in *Michael Sallis & Co Ltd* v. *ECA Calil and William F. Newman & Associates* (1987). Although the judgment was questioned in the Pacific Associates case, it would not be surprising if, at some point in the future, actions were brought by contractors against architects based on reliance on negligent specifications and perhaps contract administration.

It is important to note that, as far as the contractor is concerned, the architect's authority is stated in the contract, and is neither more nor less. Thus if, for example, the architect attempts to issue an instruction not empowered by the contract, the contractor need not carry it out. Indeed, if the contractor does carry out an instruction which is not empowered by the contract, the employer probably has no liability (but see section 4.1.3). If the architect's instruction is empowered, the contractor need not worry whether the architect has the employer's consent: the contractor may carry it out and the employer is bound.

Two simple examples should make the position clear. Assume that the architect's contract with the employer is on the terms set out in SFA/99. Paragraph 2.7 of the conditions provides that the architect is not to make any material alteration, addition or omission from the approved design without the employer's consent unless it becomes necessary during construction and the architect informs the employer without delay – in other words, if an emergency arises. Paragraph 1.6 provides that the architect must notify the employer if the services, the fees or any other part of the appointment or any information or approvals need to be varied. If, as a first example, the architect issues an instruction to vary the quality of the electrical fittings without the employer's consent, the architect will be in breach of the terms of engagement with the employer. The contractor can carry out the work, because the architect is empowered to issue such instructions by IC, and the employer must pay if any extra cost is involved. The employer can recover any costs from the architect.

As a second example, suppose the architect informs the contractor that it may take possession of the site two weeks before the appointed date. The contract gives the architect no authority to vary its terms; indeed, only the parties to the contract can do that. The contractor would take possession at its peril, because, as a party to the contract, it should be aware that the instruction is not empowered and the employer would have no liability. Whether the contractor could realistically take legal action against the architect in tort in such circumstances is probably open to debate.

If the architect is on the staff of the employer (for example, in a local authority), the position is much less clear. The contractor may assume that the architect is acting as agent for the employer. That the architect is merely acting as would any independent architect without special agency powers should be made clear to the contractor from the outset. Architects and employers in the close relationship known in law as 'master and servant' must take special care to keep their conduct above reproach. There is a duty on the employer to ensure that the architect properly carries out important duties: *Perini Corporation* v. *Commonwealth of Australia* (1969). In this respect, the contractor need have no regard to whatever standing orders may say unless they have been specifically drawn to its attention, but it may rely on the architect's apparent authority.

4.1.2 Express provisions

Article 3 of IC provides for the insertion of the name of the architect. The person whose name is entered is then the person to whom the contract refers whenever the word 'architect' appears in the conditions. Ideally, the name of the person who is actually to administer the contract should be entered. Problems can arise, however, because that person may leave the practice, die or retire. Therefore it is generally accepted that the name to be entered will be the name of the practice (i.e. XYZ & Partners) or of the chief architect in a local authority (e.g. C. Wren). Most employers and contractors accept this convention and the fact that it will be one of the partners or employees of the practice, or one of the staff of the local authority architect, who will administer the contract. An architect in private practice or local government would be wise to ensure that all interested parties, including the employer and the contractor, are informed at the beginning of the contract who are the authorised representatives (Figure 4.1). They should also be informed whenever there is a change. The situation is particularly delicate when a client has commissioned the architect, as one of the partners in a practice, to carry out the project. The client will expect that the partner personally will oversee every detail. The letter will then serve the purpose of giving assurance that, although the practice name is on the contract, the partner is personally looking after the job. The employer is under a common law duty to appoint another architect if the named architect ceases to be able to act for any reason (*Croudace Ltd* v. *London Borough of Lambeth* (1986)), and is contractually obliged to do so under article 3 and clause 3.4. All letters, instructions, certificates, notices and letters must be signed by the architect duly authorised 'for and on behalf of . . .'. It is not usually sufficient for a person to sign his or her name only, even though using headed stationery. The letter may be deemed to be written on behalf of the signatory, which may be acceptable if a partner is signing, but not so good if the signatory is merely in salaried employment. The common practice of signing in the name of another, perhaps even the architectural practice, and adding initials is acceptable provided the person signing is authorised to do so: *London County Council* v. *Vitamins Ltd* (1955). The unfortunate habit of using a rubber stamp should be avoided.

Figure 4.1
Architect to contractor, naming authorised representatives

Dear

This is to give you formal notice that the architect's authorised representatives for all the purposes of the contract are:

[*insert name*]: Partner/Director in charge of the project.
[*insert name*]: Project architect.
[*add names as required*]

The above mentioned are the only persons authorised to act as architect in connection with this contract until further written notice.

Yours faithfully

for and on behalf of [*insert the name in the contract, which should be the name of the firm of architects*].

Copy: Employer
Quantity surveyor
Consultants
Clerk of works

If the architect named in article 3 dies, or if the appointment is terminated, the employer has 14 days in which to nominate a successor (clause 3.4). It is now clear that the employer must name a successor. Except where the architect is an official of a local authority, the contractor has the right to object to the nomination (Figure 4.2). This provision is inserted because the new architect may be someone with whom the contractor has had unsatisfactory dealings in the past. If the employer thinks the contractor's reasons are insufficient, the matter can be decided by arbitration. Clearly, such a situation should be avoided if at all possible, not least because of the delaying effect on the contract.

An important provision in this article states that no succeeding architect may disregard or overrule any certificate or instruction given by the previous architect unless the previous architect would have had the power to do so. In the absence of such an express provision, no doubt a similar provision would be implied, because it is essential that the contractor's interests be safeguarded in circumstances which are solely under the control of the employer. The successor architect who disagrees with previous decisions would be wise to inform the employer, in writing, of the position, but the successor architect cannot alter them (Figure 4.3). If the contractor considers that the successor architect is attempting to disregard or overrule a previous decision, it must register its objection immediately (Figure 4.4).

The provision cannot mean that certificates or instructions given by the former architect are never, under any circumstances, to be altered. The provision means that if the successor architect considers it appropriate to make changes, they will be treated as variations and the contractor will be entitled to payment accordingly. For example, the new architect can issue instructions correcting an unsatisfactory specification and the contractor will be paid for correcting the work. Certificates have been held to be expressions of the architect's opinion in tangible form for the purposes specified in the contract: *Token Construction Co Ltd* v. *Charlton Estates Ltd* (1973). Certificates of practical completion and making good, in any event, are not appropriate to change. On the other hand, an architect effectively amends a previous financial certificate at each time for certification and, being cumulative, it is difficult to see how a new architect could forfeit this right.

Certain actions under the contract are left to the architect's discretion, for example whether to consent to the removal of unfixed goods or materials from the Works (clause 2.17). The architect will be deemed to have exercised a discretion whether or not consenting.

Figure 4.2
Contractor to employer, objecting to nomination of a replacement architect

Dear

Under the provisions of clause 3.4.1 we hereby formally give notice of our objection to the nomination of [*insert name*] of [*insert address*] as architect for the purpose of this contract in succession to [*insert name and address of previous architect*].

The grounds for our objection are [*insert particular reasons for objecting*].

A good working relationship between architect and contractor is vital to the successful completion of any project. With this in mind, we look forward to hearing that you have reconsidered the nomination.

Yours faithfully

Figure 4.3
Architect to employer if disagreement with previous architect's decisions

Dear

As you are aware, clause 3.4.2 prohibits me from disregarding or overruling any certificate or instruction given by the previous architect. This is a necessary provision to safeguard the contractor's position so far as payment for any change is concerned.

However, I must put on record that, had I been the architect at the time, I would not have [*insert as appropriate, e.g, issued instruction no 20*], and of course I take no responsibility for the consequences.

[*Add if appropriate:*]

The matter cannot be allowed to stand as it is and I propose [*insert proposals*]. I should be pleased to receive your agreement to the issue of an instruction to correct the problem in accordance with the following:

[*Insert the wording of the proposed instruction*]

Yours faithfully

Figure 4.4
Contractor to architect if ignoring a previous decision

Dear

We are in receipt of your letter/instruction [*delete as appropriate*] of the [*insert date*] in which you [*summarise contents*].

We draw your attention to the letter/instruction [*delete as appropriate*] of the [*insert date*] which we received from the previous architect. Clause 3.4.2 of the conditions of contract states that you are not entitled to disregard or overrule any certificate, opinion, decision or instruction given by the previous architect. Your letter/instruction [*delete as appropriate*] changes the requirements of the previous architect and, therefore, acts as a variation for which we are entitled to payment under the provisions of the contract. We should be grateful for your confirmation that we have correctly interpreted the position.

Yours faithfully

In exercising such discretion, the architect does not have to account for such actions to the contractor, but may have to account for them to the employer if things go wrong. So the architect should err on the side of caution.

Other clauses call for the contractor to obtain the architect's consent to certain actions, for example the sub-letting of part of the Works (clause 3.5). Figure 4.5 is a suggested letter. It is stipulated that the architect's consent must not be unreasonably delayed or withheld. If the contractor is not satisfied with the architect's decision, it can refer the matter to adjudication. The architect probably has an obligation to state reasons for withholding consent. If the architect does not give reasons, the contractor may well refer the matter to adjudication anyway. In stating such reasons, the architect should be brief.

Matters reserved for opinion – for example, whether the completion of the Works is likely to be delayed (clause 2.19.1) – are for the architect's opinion alone. The responsibility cannot be shifted onto the quantity surveyor, nor must the architect accept any interference by the employer. Such opinion must not be a whim, however, and the architect would be prudent to make file notes in case such opinion comes up for review during adjudication or arbitration.

4.1.3 The issue of instructions: general

Clause 3.8 states that all instructions which the architect issues to the contractor must be in writing. An instruction need not be written on a specially printed form headed 'Architect's Instruction', although such forms are useful for collecting all instructions in one place. Even when they are used, a careful look through the files will often unearth instructions buried within a letter. A letter is quite acceptable as an instruction, as is a handwritten instruction given on site, provided it is signed and dated. Instructions written on pieces of plywood or roofing tile used to be quite common. This author has seen an instruction written on an internal wall of a building to be reconstructed. Such practices are not to be recommended, particularly as the wall in question was demolished immediately the architect left site without any record being taken of the rather complex drawing inscribed on it. Instructions contained in site-meeting minutes are valid if the architect produces the minutes and at a subsequent meeting they are recorded as agreed. However, there would be an appreciable time lapse before such an instruction could be considered effective.

Figure 4.5
Contractor to architect, requesting consent to sub-letting part of the Works

Dear

It is our wish to sub-let [*insert details*] to [*insert name of sub-contractor*]. We do not wish to comply entirely with clause 3.6. In particular we wish to [*insert the parts it is desired to omit or change*].

In these circumstances, clause 3.5 states that we must seek your consent. We note that you are not to unreasonably delay or withhold such consent, and we look forward to receiving your response.

Yours faithfully

The position with regard to drawings is less certain. A drawing issued with a letter referring to its use on site is certainly an instruction, but a drawing issued with a compliments slip may be an instruction or it may simply be sent for comment. The contractor should make sure before carrying out the work shown thereon (Figure 4.6). If the architect simply sends a copy of the employer's letter requesting that something be done, under cover of a compliments slip, it is not an instruction but merely an invitation to the contractor to carry out the work at its own cost.

There is no provision for oral instructions. If the architect gives an oral instruction, the contractor can disregard it with impunity. If the architect confirms an oral instruction in writing, the instruction becomes effective only when the contractor receives the written confirmation. If the contractor confirms the architect's oral instruction itself, the contractor may be estopped (prevented) from contending that the instruction was not properly issued: *Bowmer & Kirkland Ltd* v. *Wilson Bowden Properties Ltd* (1996). If the employer is in the habit of issuing oral instructions, the reverse applies and the employer would be estopped from arguing that the contractor was not entitled to payment: *Ministry of Defence* v. *Scott Wilson Kirkpatrick and Dean & Dyball Construction* (2000). Oral instructions should be avoided.

The architect's authority to issue instructions is limited to those instructions of which the conditions empower issue. A list of those instructions is given in Table 4.2, and they are discussed in more detail later. The flowchart in Figure 4.7 sets out the procedure.

The contractor must comply with the architect's instructions forthwith. (This does not mean straightaway, but simply as soon as reasonably can be managed: *London Borough of Hillingdon* v. *Cutler* (1967).) The exception is an instruction requiring a variation (addition, alteration or omission) of any obligation or restriction imposed by the employer in the specification/work schedules/bills of quantities in regard to access, limitation of working space, limitation of working hours, or the order of execution or completion of work (clause 5.1.2). The contractor need not comply with a clause 5.1.2 instruction if it makes a reasonable objection in writing. Any dispute as to the reasonableness of the objection must be referred to adjudication or arbitration.

Under clause 3.10, the contractor is entitled to require the architect to specify in writing the clause that empowers the instruction (Figure 4.8). The architect must respond immediately. It is good practice to specify the empowering clause in the instruction itself. When the contractor receives the architect's response, it may do one of two things. It may carry out the instruction, in which case the clause

Figure 4.6
Contractor to architect if drawing issued without an instruction

Dear

We have today received [*insert number*] copies of your drawings numbers [*insert drawing numbers*]. There was no letter or instruction enclosed with them, merely a compliments slip.

The drawings appear to be revised versions of drawings currently in our possession/entirely new to us [*delete as appropriate*], and we should be pleased if you would instruct us whether you are issuing these drawings under the provisions of clause 2.11.

For the avoidance of doubt, please issue further drawings, details and schedules under cover of a letter making it clear that you wish us to treat the drawing as an instruction, if indeed that is your intention.

Yours faithfully

Table 4.2
Instructions empowered by IC and ICD (includes instructions noted in the schedules)

Clause	Instruction
2.1.2	Direct the integration of the CDP with the rest of the design (ICD only)
2.9	That setting out errors are not to be amended
2.11	As are necessary to enable the contractor to carry out and complete the Works in accordance with the contract (including instructions regarding the expenditure of provisional sums)
2.13.1	Correcting any departure, error or omission in the contract bills. Correcting a description in the contract bills to provide information relevant to provisional sums for defined work. Correcting any error in description or quantity, any omission or inconsistency in or between the contract documents, instructions or further issued drawings, etc. and (in the case of ICD) CDP documents
2.15.2	In regard to divergence with statutory requirements
2.30	That defective work is not to be made good
3.8	General power
3.11	Requiring a variation
3.12	Postponing work
3.14	Regarding opening up and testing
3.15.1	Regarding opening up and testing following failure of work
3.15.2	Modify clause 3.15.1 instruction
3.16.1	In regard to removal of work or materials not in accordance with the contract
3.16.2	Which are reasonably necessary if work not carried out in a proper and workmanlike manner
3.17	Excluding from site any person employed thereon
6.5.1	Take out insurance against employer's liabilities
Schedule 2 para 2	In regard to named sub-contractors: .1 to change particulars to remove impediment .2 to omit work .3 to omit work and substitute a provisional sum
Schedule 2 para 4	That work is to be carried out by person other than named in the contract documents
Schedule 2 para 7	As necessary after termination of employment of named sub-contractor: .1 to name another sub-contractor .2 to require the contractor to make its own arrangements .3 to omit the work

Figure 4.7
Architect's instructions (clause 3.8)

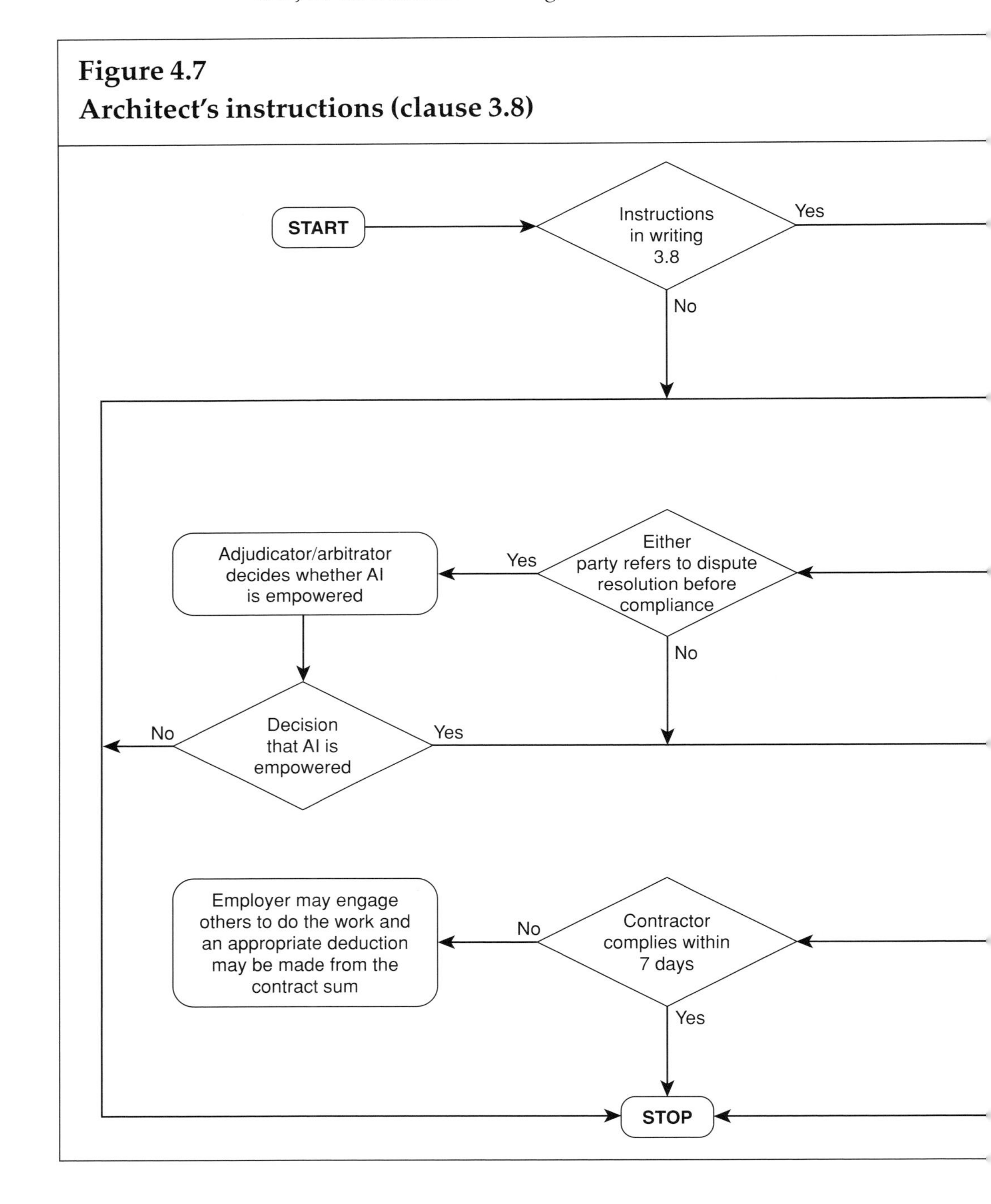

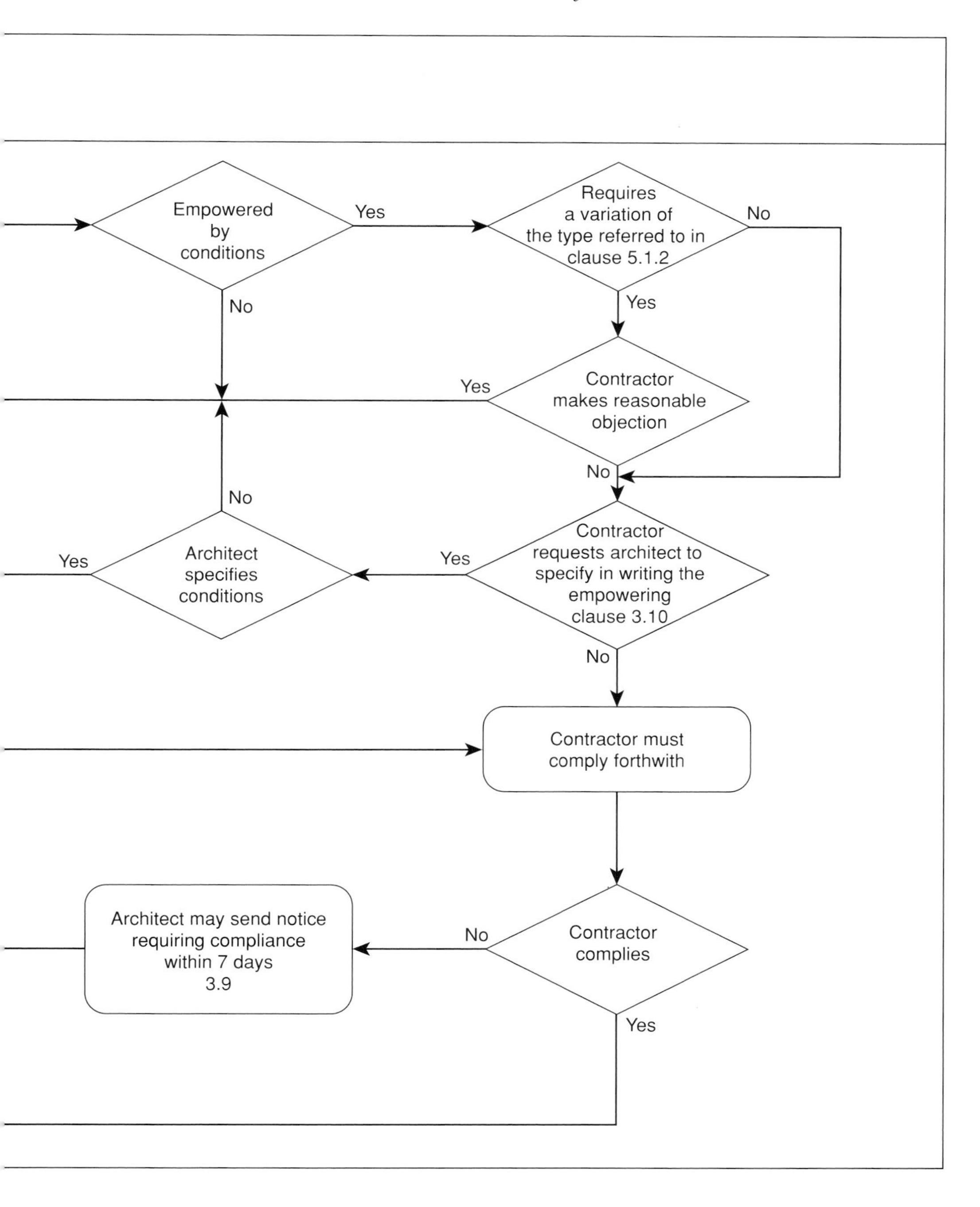
Empowered by conditions
Yes
No
Requires a variation of the type referred to in clause 5.1.2
No
Yes
Yes
Contractor makes reasonable objection
No
No
Yes
Architect specifies conditions
Yes
Contractor requests architect to specify in writing the empowering clause 3.10
No
Contractor must comply forthwith
Architect may send notice requiring compliance within 7 days 3.9
No
Contractor complies
Yes

Figure 4.8
Contractor to architect, requiring specification of clause empowering an instruction

Dear

We have today received your instruction number [*insert number*], dated [*insert date*], requiring us to [*insert a summary of the instruction*].

We request you, in accordance with clause 3.10 of the conditions of contract, to specify in writing the provision which empowers you to issue this instruction.

Yours faithfully

nominated will be deemed to be the empowering clause (whether it is or not) for all the purposes of the contract. Alternatively the contractor may invoke the dispute resolution procedures to decide whether the clause nominated does indeed empower the issue of the instruction. (The employer may give notice of adjudication or arbitration, provided that is done before the contractor has complied with the instruction.) It is thought that the contractor has the right to await the outcome of any dispute resolution procedure (adjudication or arbitration) before complying. However, the financial consequences of waiting may be heavy if the instruction is ultimately held to be properly issued. If the contractor does comply, it should write to preserve its rights (Figure 4.9).

The purpose of this provision is to protect the contractor's right to be paid if the instruction is really one that the architect is not empowered to give. Provided the architect specifies a clause – any clause – and the contractor then proceeds with the work, that clause will be the empowering clause 'under the specified provision'.

If the contractor does not carry out the architect's instruction immediately, the architect may send a written notice under clause 3.9 requiring the contractor to comply with it (Figure 4.10). If the contractor does not comply within seven days of receiving the notice, the architect should advise the employer of the right to employ and pay others to do the work detailed in the instruction, including any work that it is necessary to carry out in order to comply with the instructions. Such additional work may consist of protective work, erecting scaffolding, cutting out and reinstating. If the employer decides to take this course of action, the architect will be expected to handle the details. Wherever possible, the architect should obtain competitive quotations so that it can be shown, if it becomes necessary, that the work has been done at the lowest price it was reasonable to accept in the circumstances, in accordance with the principle of mitigation of loss. When the work is completed, what is referred to as an 'appropriate deduction' is to be made from the contract sum. The deduction will be the entire additional cost to the employer. This would include any relevant professional fees together with any additional incidental expenses caused by the contractor's non-compliance. Note that the amount deductible is the additional cost, i.e. the difference between what the work would have cost had the contractor carried out the instruction and what it did cost in fact. If the contractor, for its part, considers that the compliance notice is unreasonable, it should respond immediately (Figure 4.11). If, despite the letter, the employer persists in employing others to carry out the work a further letter from the contractor is indicated (Figure 4.12).

Figure 4.9
Contractor to architect if complying with instruction pending resolution under the relevant procedures

Dear

We refer to your letter of the [*insert date*] purporting to specify the clause empowering the issue of instruction number [*insert number*], dated [*insert date*].

As you will know, we dispute your contention and we have notified the employer that we require the dispute to be resolved by adjudication. It is our view that we are not obliged to comply with such instruction until after the adjudicator has decided whether the provision you specified empowers the issue of such instruction.

Notwithstanding the above, we are prepared to comply with your purported instruction without prejudice to the outcome of the adjudication or to any or our rights and remedies under the contract or at common law arising out of or in connection with the purported instruction, such adjudication or any other proceedings whatsoever.

We will proceed as soon as we have the employer's written acceptance of our position as set out in this letter.

Yours faithfully

Copy: Employer

Figure 4.10
Architect to contractor, giving notice requiring compliance with instruction

SPECIAL DELIVERY

Dear

Take this as notice under clause 3.9 of the conditions of contract that I require you to comply with my instruction number [*insert number*] dated [*insert date*], a further copy of which is enclosed.

If within 7 days of receipt of this notice you have not complied, the employer may employ and pay other persons to execute any work whatsoever which may be necessary to give effect to the instruction. All costs incurred thereby will be deducted from the contract sum.

Yours faithfully

Copy: Employer
Quantity surveyor

Figure 4.11
Contractor to architect on receipt of unreasonable compliance notice

Dear

We have today received your notice dated [*insert date*], which you purport to issue under the provisions of clause 3.9 of the conditions of contract.

[Add either:]

It is not reasonably practicable to comply as you require within the period specified because [*insert reasons*]. You may be assured that we have not forgotten our obligations in this matter and we intend to carry out your instruction number [*insert number*], dated [*insert date*] as soon as practicable in the light of the foregoing. Therefore, we should be pleased to hear by return that you withdraw your notice requiring compliance. If we do not receive your reply by [*insert date*], we will immediately comply, but take this as formal notice that such immediate compliance will be grounds for substantial claims for extension of time/loss and/or expense/a claim for damages at common law [*delete as appropriate*].

[Or:]

We consider that we have already complied with your instruction number [*insert number*], dated [*insert date*]. Any attempt by the employer to employ other persons and/or any deduction of any amount whatsoever from the contract sum on the basis you describe will be a serious breach of contract in respect of which we will take appropriate action. Without prejudice to the foregoing, if you will immediately withdraw your notice requiring compliance, we will be happy to meet you on site to sort out what appears to be a simple misunderstanding.

Yours faithfully

Copy: Employer

Figure 4.12
Contractor to employer if employer employs others after notice requiring compliance

SPECIAL DELIVERY

Dear

We note that you have employed other persons to carry out the contents of architect's instruction number [*insert number*], dated [*insert date*]. You purport to take this action under the provisions of clause 3.9 of the conditions of contract.

We wrote to the architect, copy to you, on the [*insert date*], explaining why you were not entitled to employ others in this instance, and warning that we would take appropriate action if you did.

Be in no doubt that we take a serious view of the situation. It appears that you are in breach of contract. We formally request you within 7 days from the date of this letter to withdraw your instructions to other persons. Failure on your part to do so will result in us taking immediate action to protect our interests in whatsoever way we deem appropriate, including, if we are so advised, accepting your actions as repudiation of the contract and recovering damages.

Yours faithfully

Copy: Architect

4.1.4 Instructions in detail

The contract empowers the architect to issue instructions in a wide variety of circumstances. In some of these circumstances there is an obligation to issue an instruction. These are marked by the use of the word 'shall', meaning 'must'. The architect must be aware of the extent of his or her powers and duties in each case.

Clause 2.13: correcting inconsistencies between contract documents; correcting errors in particulars of a named person; correcting departures from the method of preparation of the contract bills

There are four important points to note:

- The architect has an obligation to issue instructions under this clause
- The obligation to instruct does not depend on any notice from the contractor, although the contractor must notify the architect under clause 2.13.3 if any inconsistencies are found
- No inconsistencies, errors, or departures such as those mentioned will vitiate (invalidate) the contract (2.13.2)
- If the architect's instruction changes the quality or quantity of the work deemed included in the contract sum (clause 4.1), or changes employer's obligations or restrictions, a variation results.

The onus is fairly and squarely on the architect and the quantity surveyor. If they make any errors in preparing the documents, the employer will have to pay. It matters not that the errors are discovered only at a late stage in the progress of the Works. This clause is admirably clear and leaves scant room for any fudging of the issue.

Clause 2.30: not to make good defects

This clause is discussed in detail in section 9.3.4. The important point to remember is that the employer's consent must be obtained before the architect issues the instruction; the contractor has no right of objection.

Paragraph 2, schedule 2: in regard to named sub-contractors: (.1) change particulars to remove impediment; (.2) omit work; (.3) omit work and substitute a provisional sum

This clause is discussed in detail in section 8.2.2. If the architect is satisfied that the particulars specified have prevented the contractor from entering into a sub-contract, an instruction must be issued under this paragraph.

Paragraph 7, schedule 2: as necessary after termination of employment of named sub-contractor: (.1) name another sub-contractor; (.2) require the contractor to make its own arrangements; (.3) omit the work

This clause is discussed in detail in section 8.2.2. The architect has an obligation to issue an instruction under this clause, subject to receiving a written notice from the contractor stating the circumstances of the termination.

Clause 3.11: requiring a variation

This is a most important clause, giving the architect power to require the contractor to carry out variations or to sanction in writing any variation carried out by the contractor without instruction. The clause states that no such instruction or sanction will vitiate the contract. That is superfluous, because no exercise of a right conferred by the contract can vitiate that same contract. Clause 5.1 goes into some detail regarding what a variation means. In general, it means what one would assume that it means, namely the alteration or modification of the design or quality or quantity of the Works shown on the drawings and described by or referred to in the specification or work schedules or bills of quantities. Difficulties may arise, if bills of quantities are not used, in deciding just what is included and of what quality and, therefore, what is and what is not a variation (see section 3.1.2). Clause 5.1.1 proceeds to spell out in detail what kinds of situation are included: the addition, omission, or substitution of any work; the alteration of kind or standard of materials to be used in the Works; the removal from site of any work carried out or materials intended for use except where they are not in accordance with the contract. This is just a development of the initial statement and requires no further comment.

Variation is also said to mean (clause 5.1.2) the imposition by the employer of any obligations or restrictions or the addition, alteration or omission of any obligations imposed by the employer in the specification or schedules of work or bills of quantities in regard to access to, or use of, any particular parts of the site or the whole site; limitations of working space; limitations of working hours; the order of execution or completion of work. At first sight, this appears to be reserving to the employer the right to impose the obligations and to vary them. There is no mention of the architect. However, clause 3.11 seems sufficient to bring clause 5.1.2 variations within the architect's powers. The exercise of the architect's power under this clause is subject to the contractor's right of reasonable objection under clause 3.8 (Figure 4.13).

Clause 3.13: to expend provisional sums

This is mandatory. The architect must instruct the contractor how provisional sums are to be spent.

Clause 2.9: that setting out errors shall not be amended and an appropriate deduction be made from the contract sum

The architect is responsible for giving the contractor accurately dimensioned drawings and for determining any levels required. The contractor is responsible for setting out the Works correctly, i.e. in accordance with the information given by the architect. The contractor must amend any errors arising from incorrect setting out and stand the cost. The architect may instruct the contractor not to amend errors arising from inaccurate setting out, but the employer's consent must be obtained, and the architect must instruct that an appropriate deduction for the errors be made from the contract sum.

The deduction of an appropriate sum will pose difficulties. Bad setting out could result in, for example, several rooms becoming a metre shorter than intended. The wording of the clause appears wide enough to cover not only the value of work and materials omitted, but also the reduction in value of the rooms to the employer – however that is to be ascertained. Figure 4.14 is an example of the kind of letter the architect might send to the contractor in these circumstances. If the contractor objects, it should make the position clear (Figure 4.15). If the result of bad setting out is to leave the employer with a building substantially larger than required, the contractor would not be entitled to any increase for the additional work and materials, but would face the possibility of a reduction to represent additional costs to the employer (such as increased rates, cleaning charges and running costs).

What if the setting out is so bad that the building encroaches on a neighbour's land? Presumably the deduction must take all the employer's costs into account, but the situation is likely to arise only if the errors are minor, because if they are anything more, demolition is indicated.

Figure 4.13
Contractor to architect, objecting to compliance with clause 5.1.2 instruction

Dear

We are in receipt of your instruction number [*insert number*], dated [*insert date*], instructing us to [*insert a brief summary of the instruction*].

Please take this as formal notice that we make reasonable objection to compliance under the provisions of clause 3.8 of the conditions of contract. The basis of our objection is [*insert the basis of the objection, e.g., 'that closure of the access off Main Road will render the delivery and erection of the long span steel beams difficult if not impossible to achieve'*].

We should be pleased, therefore, if you would withdraw your instruction or amend it to deal with our objection.

Yours faithfully

Figure 4.14
Architect to contractor, referring to setting out errors not to be amended

Dear

I refer to the meeting on site yesterday, and I confirm the position with regard to the setting out as follows:

1. At the commencement of this contract, you were provided with all necessary levels and accurately dimensioned drawings to enable you to set out the Works.

2. The inspection yesterday revealed that there were errors in the setting out for which you were responsible under the provisions of clause 2.9 of the conditions of contract. The errors are [*insert a description of the errors*], which are obviously not easily amended at this stage in the progress of the Works.

3. In accordance with clause 2.9, I enclose an architect's instruction instructing you not to amend the errors. I have requested the quantity surveyor to carry out the necessary calculations so that an appropriate deduction for such errors may be made. [*If appropriate, add: 'You should note that the deduction will include a sum which will reflect the reduction in value of the finished Works.'*]

Yours faithfully

Copy: Employer

Figure 4.15
Contractor to architect, objecting to amount of deduction after setting-out errors not amended

Dear

We are in receipt of your letter and instruction of the [*insert date*]. We note that you instruct us not to amend the setting out of [*insert details*].

We acknowledge that you are empowered to issue such an instruction, but we strongly object to a deduction of money such as you indicate in your letter. Clause 2.9 refers to a deduction. Nowhere in the contract is it stipulated how such a deduction is to be calculated; only that it is to be 'appropriate'. In our view, the amount of such a deduction can only be ascertained by agreement. No doubt you will present your proposals for our consideration in due course. If you attempt to deduct any amount whatsoever from the contract sum before we have signified our agreement in writing, we shall take immediate action for recovery.

Yours faithfully

Copy: Employer

The contractor is given no choice in the matter, but, in the face of what seems to be an unreasonable attitude on the part of the employer, the dispute is likely to be referred to adjudication or arbitration.

Clause 3.14: requiring opening up or testing

This clause empowers the architect to instruct the contractor to open up for inspection any work which has been covered up, or to arrange for any testing of materials whether or not they are already built in. The architect will instruct the contractor to open up or test because it is suspected that work or materials are defective. There may be no alternative, but the position is that, if the work or materials are found to be in accordance with the contract, the cost (including the cost of making good) must be added to the contract sum unless provision is expressly made for opening up or testing in the specification or work schedules or bills of quantities. If the work or materials are found to be not in accordance with the contract, all the costs must be borne by the contractor.

Clause 3.15: requiring opening up or testing at contractor's cost if similar work or materials have failed

This clause gives the architect a useful power to check for possible defective work. The exercise of that power will give rise to further responsibilities. Before the architect may issue an instruction, work or materials must have been discovered which are not in accordance with the contract. After the discovery, the contractor must write to the architect stating what it intends to do immediately to ensure that there is no similar failure in the work already carried out or materials already supplied. The proposals must be at no cost to the employer.

An important point is that only similar failures are under consideration. For example, if it is discovered that, in one section of the work, wall ties have not been provided in sufficient numbers, the architect will expect to be satisfied that wall ties have been properly provided in other parts of the building. Thus the contractor will not be concerned, on that occasion, with possible failures of damp-proof courses or foundations. It is debatable whether the contractor must also satisfy the architect, in this example, that the other wall ties have been properly bedded, if that was not the reason for failure. On balance, there is probably no such duty. In most cases, of course, the contractor's proposals will give the architect the opportunity of checking that some other defects are not present.

Since, in the course of a contract, there will be many instances of failures, large and small, the architect should make sure that the contractor is aware of which instances are considered significant enough to give rise to proposals under this clause (Figure 4.16). But it must be stressed that the onus is on the contractor, and the architect has no contractual duty to make the contractor aware. Power to issue an instruction arises after discovery of failure if the architect:

- has not received the contractor's proposals within seven days of discovering the failure; or
- is not satisfied with the action proposed by the contractor; or
- cannot wait for the contractor's written proposals because of safety considerations or statutory obligations.

The architect may then issue an instruction which requires the contractor to open up for inspection or to arrange for testing any work or materials, whether built in or not. The instruction should state that the work will be at no cost to the employer. The whole process is at the contractor's own cost, whether or not the opening up or testing discovers further failures. (This is in contrast to the provisions of clause 3.14.) The contractor must forthwith comply with the architect's instruction. If not, the architect may obtain the employer's consent to applying clause 3.9 remedies (see section 4.1.3).

Clause 3.15.2 gives the contractor 10 days from receipt of the instruction to decide whether to object to compliance. The right of objection is stated to be without affecting the obligation to comply. This means that the contractor must still comply even though objecting. If the contractor decides to object, it must inform the architect in writing, stating the reasons (Figure 4.17). The architect has seven days from receipt of the objection to withdraw the instruction or to modify the instruction to take care of the contractor's objection.

If the architect takes neither action, any dispute regarding whether the nature or extent of the opening up or testing was reasonable in all the circumstances is to be referred to adjudication or arbitration.

The dispute resolution procedure is to decide whether the instruction is reasonable in all the circumstances. If it is decided that it is not, the adjudicator or arbitrator as the case may be must decide what amount the employer must pay to the contractor for carrying out the work, including making good. It is clearly envisaged that it may be held that the instruction is reasonable in essence but that the architect has gone too far in requiring existing work to be opened up. In such a case, the adjudicator or the arbitrator has power to order the employer to pay a contribution. The contractor, however, must

Figure 4.16
Architect to contractor, following failure of work

Dear

When I visited site today, I noted that [*specify work or material*] were not in accordance with the contract.

Under the provisions of clause 3.15.1 of the conditions of contract, I require you to state in writing, within 7 days of the date of this letter, what action you will immediately take at no cost to the employer to establish that there is no similar failure to work already executed / materials or goods already supplied [*omit as appropriate*].

Yours faithfully

Figure 4.17
Contractor to architect, objecting to compliance with clause 3.15.1 instruction

SPECIAL DELIVERY

Dear

We are in receipt of your instruction number [*insert number*], dated [*insert date*], instructing us to [*insert nature of the work*], which you purport to issue under clause 3.15.1 of the conditions of contract. We consider such instruction unreasonable because [*state reasons*].

If within 7 days of receipt of this letter you do not in writing withdraw the instruction or modify it to remove our objection, a dispute or difference will exist about whether the nature or extent of the opening up/testing [*delete as appropriate*] in your instruction is reasonable in all the circumstances, such dispute or difference to be referred to immediate adjudication in the first instance. In such event, we will comply with our obligations pending the result of such adjudication and award of additional costs and extension of time.

Yours faithfully

Copy: Employer

still comply with the instruction. In most cases the contractor is unlikely to wish to invoke the dispute resolution procedures, but the wording of the clause makes such procedures inevitable if the architect ignores any objection the contractor makes. The contractor may take advantage of that to try and force concessions when making objection. In practice, the dispute resolution procedures will not take place against the wishes of both parties. It will be seen that, if all parties take the maximum allowable time to act under the various parts of the clause, some 26 days will have elapsed between the discovery of the failure and the date on which dispute resolution becomes automatic. The procedures and options in this clause are set out in the flowchart in Figure 4.18.

Clause 3.16: to remove defective work from the site

This clause empowers the architect to order that defective work or materials be removed from site. Defective work or materials are work or materials not in accordance with the contract. The architect has no power to simply order defective work or materials to be rectified (even under clause 2.30, the architect may only notify the contractor of defects). To be effective, the instruction must order removal from site: *Holland Hannen and Cubitts (Northern)* v. *Welsh Health Technical Services Organisation* (1981). In most cases it is to be expected that the contractor will correct defective work or materials without the necessity for an instruction, but the architect cannot put any sanction into operation until removal has been instructed and the contractor has not complied. So the architect should always send an instruction under this clause whenever work or materials are found to be defective. For the same reason, the contractor must not hesitate to put the matter on record if it thinks the instruction is not justified (Figure 4.19).

Clause 3.12: postponing work

The architect is entitled to issue an instruction to postpone any of the work required by the contract. There is a price to pay, and the employer has to pay it, so the architect must take care. If an instruction is given to postpone work, the contractor is entitled in principle to an extension of time (clause 2.20.2.1) and loss and/or expense (clause 4.18.2.1). Under clause 8.9.2.1 the contractor may terminate its employment if the whole or substantially the whole of the Works are suspended for a continuous period of the length (usually 2 months) stated in the contract particulars due to postponement, among other things (see section 12.2.2). It has been held that, in certain circumstances, an instruction given on another matter may imply postponement, with all its consequences: *M. Harrison & Co Ltd* v. *Leeds City Council* (1980). The situation should not arise if the architect is careful to quote the correct empowering clause in each case and use the wording of the clause as far as appropriate.

Clause 3.17: to exclude persons from the Works

This short clause has been introduced to place beyond doubt that the architect has power to give instructions requiring exclusion of persons from site. The instruction must not be issued unreasonably or vexatiously. In other words, the architect must have a good reason to instruct the exclusion. In practice, this will probably be as a result of consistent bad workmanship, but bad behaviour could also qualify.

4.2 *Duties*

4.2.1 Duties under the contract

In section 4.1.1 the architect's duties under the contract were seen to flow directly from the terms of engagement with the employer. Many of those duties can be recognised because they are preceded by the word 'shall', e.g. 'the architect shall . . .'. A full list is contained in Table 4.1.

In performing such duties, architects are expected to act competently; they are expected to act with the same degree of skill and care as the average competent architect: *Bolam* v. *Friern Hospital Management Committee* (1957). If architects profess greater than average skill, either generally or in some special area, that is the standard by which they will be judged.

Clause 2.2, within the general clause about contractor's obligations, paradoxically places a potentially heavy duty on the architect. In essence, it states that if approval of workmanship or materials is a matter for the opinion of the architect, the quality and standards must be to the architect's reasonable satisfaction. At first sight, this clause appears to give the architect considerable power, and so it does. The catch is contained in clause 1.10.1.1, which states that the issue of the final certificate is conclusive evidence that where anything is described expressly in any drawings, bills of quantities, specification, work schedules or instructions to be for the architect's

Figure 4.18
Failure of work (clause 3.15)

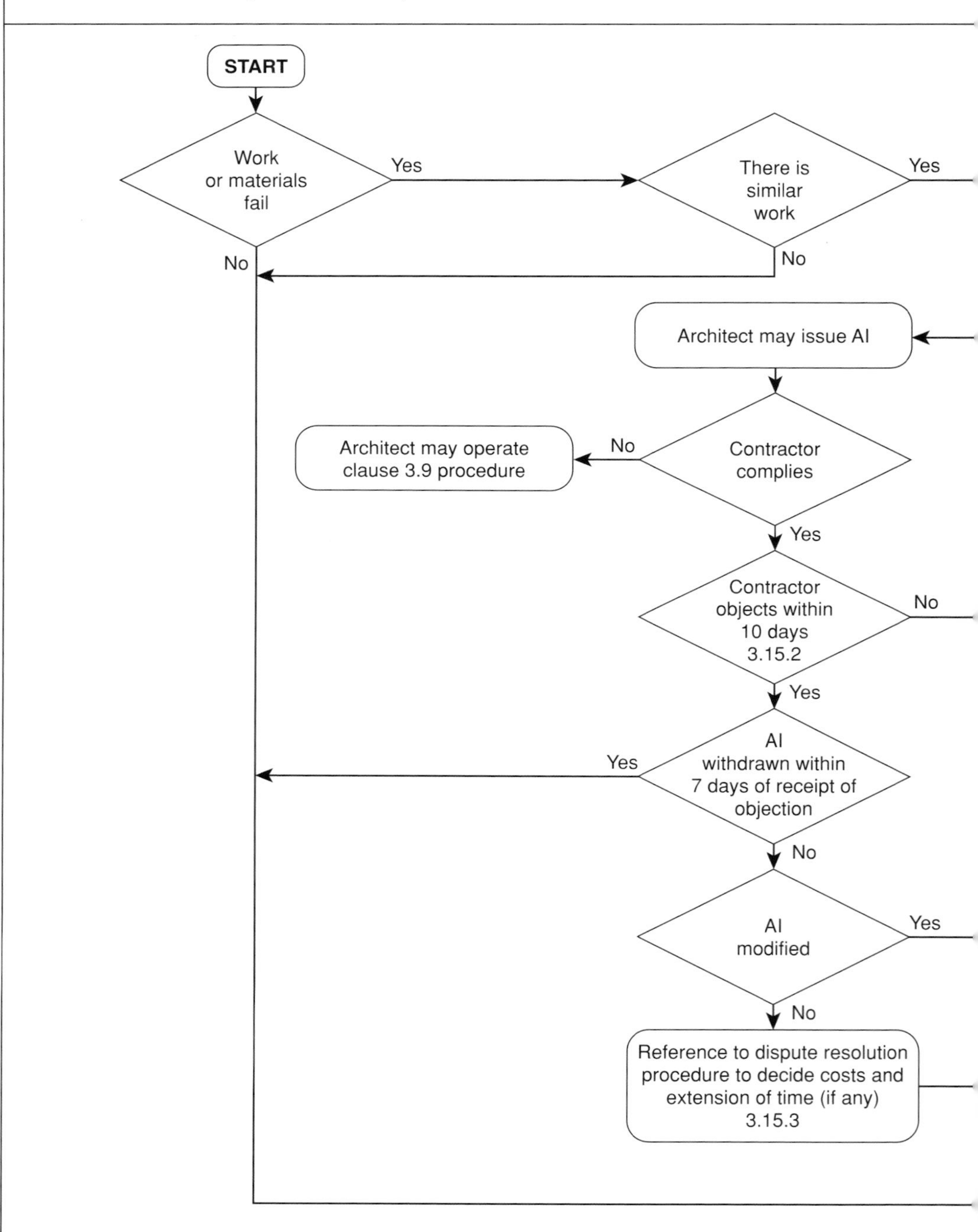

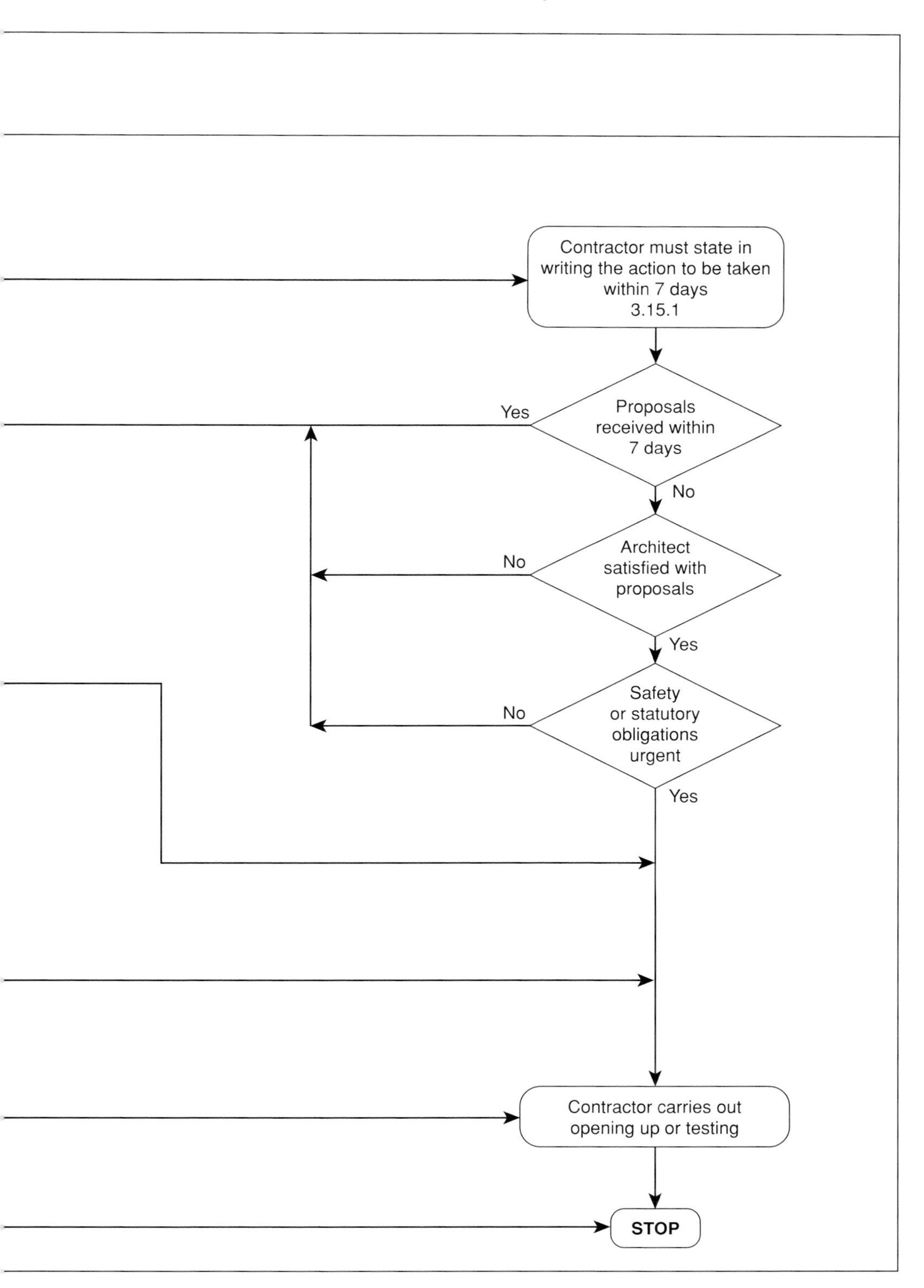
Contractor must state in writing the action to be taken within 7 days 3.15.1
Proposals received within 7 days
Yes
No
Architect satisfied with proposals
No
Yes
Safety or statutory obligations urgent
No
Yes
Contractor carries out opening up or testing
STOP

Figure 4.19
Contractor to architect if instruction to remove defective work is not justified

Dear

We are in receipt of your instruction number [*insert number*], dated [*insert date*], instructing us to remove [*specify the work noted*] from site.

You purport to issue your instruction under the provisions of clause 3.16 of the conditions of contract. Clause 3.16 expressly restricts the issue of such instruction to work, materials or goods which are not in accordance with the contract.

We formally give notice that the work/material/goods [*delete as appropriate*] noted in your instruction is in accordance with the contract and that, therefore, your instruction is invalid under clause 3.16.

If you withdraw your instruction and rephrase it so as to bring it under the provisions of clause 3.11,we shall be happy to comply.

Yours faithfully

approval, it is to the architect's reasonable satisfaction. Architects commonly specify that various items are to be to their approval, sometimes even going so far as to state generally that 'unless otherwise stated, all workmanship and materials are to be to the approval of the architect'.

The combined effect of clauses 2.2 and 1.10.1.1 is that, if anything is reserved for the architect's approval, the final certificate confirms that it is to the architect's reasonable satisfaction. That is the case whether or not the architect has specifically expressed approval during the course of the work. Therefore the architect who puts a general approval clause in the contract documents takes on a duty of approving everything to which it applies. If anything is missed, the final certificate will make it to the architect's reasonable satisfaction. Moreover, if the architect approves something which is not in accordance with the contract, such approval will usually override the contract requirements. It cannot be said later that it was not in accordance with the contract and therefore unacceptable. The architect should, therefore, severely limit the items to which the right of approval is reserved.

The case of *Colbart Ltd* v. *H. Kumar* (1992) and the subsequent Court of Appeal case *Crown Estates Commissioners* v. *John Mowlem & Co Ltd* (1994) caused something akin to panic among architects by holding that all matters of standards of workmanship and quality of materials were matters for the architect's opinion. Effectively, that made the final certificate conclusive about the architect's satisfaction with the whole of the work and materials. The current clause 1.10.1.1, as noted above, modifies the position and restricts the conclusivity of the final certificate to the matters expressly reserved in the contract documents. Moreover, by a final phrase, the final certificate is stated not to be conclusive that any goods or materials comply with any other requirement of the contract. This leaves the door open for the employer to seek recourse from the contractor, even where the architect's satisfaction has been conclusively established. Table 4.3 lists the certificates to be issued by the architect.

4.2.2 General duties

This is not the place to discuss in detail architects' general duties to their clients and to third parties, but one aspect of their general duties affects the contract. They have a duty to their client to be familiar with those parts of the law which affect their work. For example, clients will expect architects to have a thorough knowledge

Table 4.3
Certificates to be issued by the architect under IC and ICD

Clause	Certificate
2.21	Practical completion
2.22	Non-completion
2.31	Making good
4.6.1	Interim
4.9	Interim on practical completion
4.14.1	Final

of the planning laws. This is not the specialist knowledge expected of a lawyer or planning consultant who deals with nothing but planning appeals, but is the knowledge which architects require to advise their clients and make successful planning applications on their behalf. Similarly, the architect is expected to have a thorough knowledge of the various forms of contract, so that they can advise their clients which contract is most appropriate for a particular job. Architects must be knowledgeable about the particular contract which they advise their clients to use.

Architects must be aware of the pitfalls in a contract and the correct interpretation of each clause as shown by any applicable case law. If an architect has never read a contract, much less a legal commentary, then there will be a serious problem, and the architect will be unable to carry out duties under the contract or to act fairly within the meaning of the contract. Such an architect will be unable to advise a client when a particularly difficult contractual point arises; and there are likely to be more mistakes than occur with the average competent architect.

If a client is put to unnecessary expense by the architect's inadequate knowledge of contractual provisions, there may well be an action for damages suffered as a result of the architect's incompetence.

4.3 Summary

Authority

- An architect's authority depends on what has been agreed with the employer
- Only the employer can take action against the architect in contract
- Employer and contractor may be able to take action against the architect in tort
- The architect has a dual role once the contract is signed
- Architects are not quasi-arbitrators; nor are they immune from actions for negligence
- The contractor is entitled to see the architect's authority only in terms of the contract
- Only the architect named in the contract or his or her authorised representative can exercise any powers under the contract
- If one architect takes over a contract from another, decisions of the previous architect cannot be altered without the contractor being entitled to payment, but the employer should be informed

- Matters reserved for the architect to decide cannot be delegated to another (e.g. the quantity surveyor)
- Instructions must be clear and in writing
- The architect may issue only those instructions of which the contract specifically empowers the issue
- The architect must be aware of specific powers and duties in the case of each instruction.

Duties

- Architects must be as skilled as the average architect
- Architects will be judged on any skill which they profess to possess over and above the average
- If items are stated to be to the architect's approval, there is a duty to approve each one
- Architects have a duty to know the law as it affects the performance of their profession
- Architects must know the law as it affects the contract.

CHAPTER FIVE
THE CONTRACTOR'S OBLIGATIONS

5.1 *Express and implied obligations*

5.1.1 Legal principles

Apart from the express terms of the contract, the general law requires that the contractor will do three things:

- The contractor must carry out its work in a good and workmanlike manner, i.e. show the same degree of competence as the average contractor experienced in carrying out that type of work
- The contractor will supply good and proper materials
- The contractor must complete the work by the date for completion stated in the contract or, if no date is specified, within a 'reasonable time' of being given possession of the site.

In addition, if there was no designer employed, the contractor would be obliged to ensure that the finished building was reasonably fit for its intended purpose so far as it had been made known.

These obligations may be modified by the terms of the contract itself. Under the general law, the contractor is also responsible to the employer for the work done, and goods and materials provided, by its sub-contractors and, in common with most standard form building contracts, under IC the contractor is responsible to the employer for all the defaults of a sub-contractor, whether named or otherwise. But in the case of named sub-contractors (see section 8.2.2) this liability is substantially modified by the very wide terms of paragraph 11.1 of schedule 2, which exempts the contractor from responsibility to the employer for design and allied failures in the named sub-contractor's work. This apart, the position under IC is that the main contractor is liable to the employer for all other sub-contractors' defaults, of fabrication, workmanship or otherwise.

Statutory obligations are also imposed on contractors by the Supply of Goods and Services Act 1982, although these statutory obligations are included in the express terms of IC. The contract must be read against this background. It imposes many specific duties on the contractor, some of which alter or affect the common law position. These obligations are scattered throughout the printed form, and only the more important of them are collected in this chapter.

Table 5.1 summarises the contractor's powers and duties under the express terms of the contract.

5.1.2 Execution of the Works

Clause 2.1 requires the contractor to 'carry out and complete the Works in a proper and workmanlike manner and in compliance with the contract documents', which are defined in clause 1.1; this is a basic and absolute obligation. It is not qualified in any way. The contractor must bring the Works to a state where they are practically completed so that the architect can issue a certificate under clause 2.21. This is what the contractor must do no matter what difficulties may be encountered, but subject to what is said in clause 8.11 about the termination of the contractor's employment under the contract for causes outside the control of either party.

The work must be carried out and completed 'in compliance with' the contract documents as defined. It is the architect's responsibility to ensure that the description of the work is adequate, and care should be taken to use precise wording. Generalisations are impossible to enforce. The contract documents must contain all the requirements which the employer wishes to impose, and the use of phrases such as 'of good quality' or 'of a durable standard' should be avoided.

The contractor must complete all the work shown in, described by, or referred to in the contract documents. This obligation is ended only when the architect issues the certificate of practical completion. Thereafter the contractor must remedy defective work during and immediately after the specified rectification period (see section 9.3).

Clause 2.2.1 states that if approval of workmanship or materials is a matter for the architect's opinion, then the quality and standard must be to the architect's reasonable satisfaction. The effect of this is discussed in section 4.2.1.

The basic contractual obligation is amplified by clause 2.4. Once given possession of the site, the contractor must begin and proceed 'regularly and diligently' with the Works and complete them on or before the specified completion date, as extended. Failure to proceed 'regularly and diligently' is one of the grounds which may give rise to determination of its employment under the contract by the employer (see clause 8.4.1.2).

Table 5.1
Contractor's powers and duties under IC and ICD (including powers and duties in the schedules)

Clause	Power/duty	Precondition/comment
2.1	**Duty** Carry out and complete the Works in a proper and workmanlike manner and in accordance with the contract documents, health and safety plan and statutory requirements	
2.1.1 (ICD only)	**Duty** Complete the design for CDP	To include selection of specifications and materials, goods and workmanship so far as not in the Employer's Requirements or Contractor's Proposals
2.1.2 (ICD only)	**Duty** Comply with architect's integration directions	Subject to the provisions of clause 3.8.2
2.2.2	**Duty** Take reasonable steps to encourage contractor's persons to be registered under CSCS	
2.3	**Duty** Pay all fees and charges	Legally recoverable. Amount to be added to the contract sum unless they are to be included *or* (in the case of ICD) relate solely to CDP
2.4	**Duty** Thereupon begin construction, regularly and diligently proceed and complete on or before the relevant completion date	Subject to extension of time provisions
2.6	**Power** Consent to employer using or occupying the site or Works before the issue of practical completion certificate **Duty** Notify insurers (Options A, B or C) and obtain confirmation that insurance will not be prejudiced. Notify employer of additional premium and provide premium receipt to employer on request	If insurers confirm that insurance will not be prejudiced

Table 5.1 *Cont'd*

Clause	Power/duty	Precondition/comment
2.7	**Duty** Permit execution of work not forming part of the contract to be carried out by the employer or person directly engaged concurrent with the contract Works **Power** Consent to carrying out of such work	If contract documents so provide Where employer requests and contract documents do not so provide. Consent must not be unreasonably delayed or withheld
2.8.3	**Duty** Use contract documents and any further documents issued under the contract only for the purposes of the contract	Contractor must amend any errors arising from inaccurate setting out at no cost to employer
2.9	**Duty** Be responsible for setting out	
2.10	**Power** Agree with employer to vary time for release of information from the information release schedule	Reasonably necessary to amplify Contractor's Proposals
2.10.2 (ICD only)	**Duty** Provide architect with two copies of contractor's design documents and, if requested, calculations and information and all setting out and levels used for CPD	
2.10.3 (ICD only)	**Duty** Not to commence work until design submission procedure has been complied with	In relation to CDP work
2.11.3	**Duty** Advise the architect sufficiently in advance that drawings are needed	If the contractor has reason to believe that the architect is not aware of the time by which the contractor needs the drawings
2.13.3.1	**Duty** Immediately give written notice to the architect with details	If contractor finds departure, error, omission or inconsistency in or between documents

Table 5.1 *Cont'd*

Clause	Power/duty	Precondition/comment
2.13.3.2 (ICD only)	**Duty** As soon as practicable make written proposals for necessary amendments	If there is an inconsistency in Employer's Requirements not dealt with in Contractor's Proposals or between CDP documents other than Employer's Requirements
2.15.1	**Duty** Immediately give the architect written notice specifying the divergence **Duty** Inform the architect in writing of proposals for removing the divergence	If the contractor finds any divergence between statutory requirements, contract documents, further information and instructions. Where (in the case of ICD) the divergence involves CDP documents
2.15.2	**Duty** Comply (in the case of ICD) without cost to the employer	Unless after base date there is change in statutory requirements requiring an alteration in the CDP
2.16.1	**Duty** Supply such limited materials as are reasonably necessary to secure immediate compliance with statutory requirements	In an emergency where it is necessary to act before receiving an architect's instruction
2.16.3	**Duty** Forthwith inform the architect of such emergency compliance	
2.19.1	**Duty** Notify the architect in writing forthwith of any cause of delay	If it becomes reasonably apparent that progress of the Works is being or is likely to be delayed. The duty covers all causes of delay and is not confined to the relevant events in clause 2.20
2.19.4	**Duty** Constantly use best endeavours to prevent delay and do all reasonably required to the architect's satisfaction **Duty** Provide such information required by the architect as reasonably necessary	The contractor is not expected to spend large sums of money or there would be no need for an extension of time clause

Table 5.1 *Cont'd*

Clause	Power/duty	Precondition/comment
2.23	**Duty** Pay or allow to the employer liquidated damages at the rate specified in the contract particulars	If the Works are not completed by the specified or extended completion date; *and* If the architect has issued a certificate of non-completion; *and* If the employer has informed the contractor in writing before the date of the final certificate that payment may be required or a deduction made; *and* If the employer has required liquidated damages in writing not later than 5 days before the final date for payment
2.25	**Power** Consent to the employer taking possession of part of the Works before practical completion or section completion certificate has been issued	Consent must not be unreasonably delayed or withheld
2.30	**Duty** Make good any defects, shrinkages or other faults at no cost to the employer	If the defects, etc. appear and are notified to the contractor by the architect not later than 14 days after expiry of the rectification period; *and* If they are due to materials or workmanship not in accordance with the contract; *and* The architect has not instructed otherwise
2.32 (ICD only)	**Duty** Supply for retention and use of the employer such CDP documents and related information as may be specified or the employer may reasonably require showing the CDP as built	Before practical completion without further charge to the employer
3.2	**Duty** Ensure that there is a person in charge on the site	At all reasonable times

Table 5.1 *Cont'd*

Clause	Power/duty	Precondition/comment
3.4.1	**Power** Object to a replacement architect	Must be for reasons which will be considered sufficient by a person under the dispute resolution procedure
3.6.3	**Duty** Not to give consent under clause 3.6.2.1 without prior consent of architect	
3.7	**Duty** Enter into a sub-contract with a named person	Not later than 21 days after execution of main contract
3.8	**Duty** Forthwith comply with architect's instructions	If the architect is empowered to issue them Contractor need not comply with instructions requiring clause 5.1.2 variations if reasonable written objections given
3.8.2 (ICD only)	**Duty** Specify in writing injurious effect of direction or instruction	If direction or instruction in the contractor's opinion will injuriously affect efficacy of the design of CDP
3.10	**Power** Request the architect to specify in writing the empowering clause for an instruction	
3.15.1	**Duty** State in writing to the architect the action which the contractor proposes to take immediately to establish that there is no similar failure of work, etc. Forthwith comply with any architect's instruction requiring opening up for inspection and testing	Where such failure is discovered during the carrying out of the Works Unless within 10 days of receipt of the instruction the contractor objects to compliance, stating its reasons in writing. Then if within 7 days of receipt of the contractor's objection the architect does not withdraw or modify the instructions, the dispute or difference is referred to the relevant dispute resolution procedures

Table 5.1 *Cont'd*

Clause	Power/duty	Precondition/comment
3.16.1	**Duty** Comply with architect's instructions about the removal of work from site	
3.18.1	**Duty** Comply with the duties of the principal contractor under the CDM Regulations	
3.18.2	**Duty** Ensure that the health and safety plan is received by the employer before construction work commences. Notify the employer	If it is the principal contractor If contractor amends the plan
3.18.3	**Duty** Provide and ensure any sub-contractor provides information reasonably required for the health and safety file	Within the time reasonably required by the planning supervisor
3.19	**Duty** Comply at no cost	If a successor principal contractor is appointed
4.6.3	**Power** Submit an application to the quantity surveyor, setting out the contractor's view of the amount due	Not later than 7 days before the certification date. The quantity surveyor must make an interim valuation and, to the extent that it differs from the contractor's application, the quantity surveyor must notify the contractor, using similar detail to the application
4.11	**Power** Suspend performance of obligations	If the contractor has given 7 days' written notice of intention after the employer has failed to pay in full the amount on a certificate by the final date for payment
4.13.1	**Duty** Not later than 6 months after practical completion send to the architect (or to the QS if the architect so instructs) all documents reasonably required for the purposes of the adjustment of the contract sum	

Table 5.1 *Cont'd*

Clause	Power/duty	Precondition/comment
4.17	**Power** Make written application to the architect within a reasonable time **Duty** Submit to the architect or quantity surveyor such information as is reasonably necessary to enable an ascertainment of direct loss and/or expense to be made	If it becomes apparent that regular progress is being materially affected by one or more of the relevant matters or due to deferment of possession of the site by the employer
5.2	**Power** Agree with employer the amount of a variation instruction or instruction on the expenditure of a provisional sum	
6.1	**Duty** Indemnify the employer against any expense, liability, loss claim or proceedings whatsoever in respect of personal injury or death of any person	The claim must arise out of, or in the course of, or be caused by, the carrying out of the Works and not be due to any act or neglect of the employer or any of the employer's persons
6.2	**Duty** Similarly indemnify the employer against property damage	If the claim arises out of, or in the cause of, or is caused by reason of the carrying out of the Works *and* is due to any negligence, omission or default of the contractor or contractor's persons
6.4.1	**Duty** Maintain insurance in joint names of employer and contractor for such amounts of indemnity as are specified in the contract particulars and in compliance with all relevant legislation	
6.4.2	**Duty** Send documentary evidence of insurance to the architect for inspection by the employer	As and when reasonably required by the employer

Table 5.1 *Cont'd*

Clause	Power/duty	Precondition/comment
6.5.1	**Duty** Take out insurance in joint names in respect of any expense, liability, loss, claim or proceedings incurred by the employer for injury or damage caused by collapse, subsidence, heave, vibration, etc. by reason of carrying out the Works	If the contract particulars so state and if the architect so instructs
6.5.2	**Duty** Send to the architect for deposit with the employer, the policy and premium receipts	
6.9.1	**Duty** Ensure that joint names policy *either*: • provides for recognition of each sub-contractor as an insured; *or* • includes a waiver of subrogation in regard to such sub-contractors.	Where insurance option A applies. In respect of loss or damage by specified perils to the Works or section. Continues to practical completion or termination if earlier
6.10.1	**Duty** Inform the employer	If insurers named in joint names policy notify the contractor that terrorism cover will no longer be available
6.10.4.1	**Duty** With due diligence restore damaged work, replace or repair materials, dispose of debris and proceed with carrying out the Works	If work or materials suffer loss or damage due to terrorism and the employer has not given notice of termination
6.12	**Duty** Ensure compliance with Joint Fire Code of all contractor's persons	
6.13.1	**Duty** Send copies of notice to employer and architect	If breach of Joint Fire Code occurs and insurers notify the contractor of remedial measures required
6.13.1.1	**Duty** Ensure the remedial measures are carried out by the specified date	If the measures relate to the contractor's obligation to carry out and complete the Works

Table 5.1 *Cont'd*

Clause	Power/duty	Precondition/comment
6.13.1.2	**Duty** Supply limited materials and execute limited work as reasonably necessary **Duty** Inform the architect of the emergency and steps taken	In the case of emergency compliance with remedial measures Forthwith
6.15.1 (ICD only)	**Duty** Take out PI insurance with limit of indemnity of type and amount not less than in contract particulars	Forthwith after contract entered into
6.15.2 (ICD only)	**Duty** Maintain PI insurance until expiry of period in contract particulars from practical completion	Provided it remains available at commercially reasonably rates
6.15.3	**Duty** Produce documentary evidence of insurance	When reasonably requested by employer or architect
6.16	**Duty** Immediately give notice to employer	If insurance ceases to be available at commercially reasonable rates. Employer and contractor must discuss best means of protecting their positions
7.5	**Duty** Enter into a collateral warranty in form CWa/P&T with purchaser or tenant	Within 14 days of employer notifying the names and interests of such persons who are already identified in the contract particulars
7.6	**Duty** Enter into a collateral warranty in form CWa/F with funder	Within 14 days of employer notifying the names and interests of such person who is already identified in the contract particulars
7.7	**Duty** Comply with requirements regarding the obtaining of warranties from sub-contractors to purchasers, tenants or funder in form SCWa/P&T or SCWa/F	Within 21 days of employer giving notice identifying such persons who are already named in the contract particulars
7.8	**Duty** Comply with requirements regarding the obtaining of warranties from sub-contractors to the employer	Within 21 days of employer giving notice if the contract particulars so provide

Table 5.1 *Cont'd*

Clause	Power/duty	Precondition/comment
8.5.2	**Duty** Immediately inform the employer in writing	If contractor makes any proposal, gives notice of meeting or becomes subject of proceedings, etc. relating to clause 8.1 matters
8.7	**Power** Serve a notice on the employer by special, recorded or actual delivery specifying the default or suspension event	If the employer: Does not pay by the final date for payment an amount properly due to the contractor on any interim or final certificate; *or* Interferes with or obstructs the issue of any certificate; *or* Fails to comply with the assignment provisions; *or* Fails to comply in accordance with the contract with the CDM Regulations; *or* The carrying out of the whole or substantially the whole of the uncompleted Works is suspended for a continuous period as stated in the contract particulars by reason of • Architect's instructions under 2.13 (errors and inconsistences), 3.11 (variations) or 3.12 (postponement) • Any impediment or default of the employer, architect, quantity surveyor or employer's persons unless caused by contractor's or contractor's persons' negligence or default
8.9.3	**Power** Terminate its employment within 10 days of expiry of the 14 day period	If default or suspension event has not ceased within 14 days
8.9.4	**Power** Terminate its employment	Within a reasonable time after repetition if the employer repeats a default or suspension event for which the contractor for any reason has not already terminated

Table 5.1 *Cont'd*

Clause	Power/duty	Precondition/comment
8.10.1	**Power** Terminate its employment	If the employer is insolvent
8.11.1	**Power** Give 7 days notice of termination	If the carrying out of the whole or substantially the whole of the Works is suspended for a period stated in the contract particulars by reason of *force majeure* or architect's instructions under clauses 2.13, 3.11 or 3.12 resulting from statutory undertaker's negligence of default or loss or damage due to specified perils or civil commotion or threat of terrorism or exercise by UK government of statutory powers
	Power Give further notice of termination	If suspension has not ceased within the 7-day period
8.12.2	**Duty** Prepare an account	As soon as reasonably practicable following termination by the contractor under 8.11
9.1	**Power** Agree to resolve a dispute by mediation	
9.3	**Power** By written notice jointly with the employer to the arbitrator state that they wish the arbitration to be conducted in accordance with any amendments to the JCT 2005 CIMAR	
9.4.1	**Duty** Serve on the employer a written notice	If the contractor wishes a dispute to be resolved by arbitration
9.4.3	**Power** Give a further arbitration notice to the employer referring to any other dispute	After the arbitrator has been appointed. Rule 3.3 applies
Schedule 1 para A.1	**Duty** Take out and maintain insurance for new buildings	In joint names for all risks

Table 5.1 *Cont'd*

Clause	Power/duty	Precondition/comment
Schedule 1 para A.2	**Duty** Send the para A.1 policy to the architect	For deposit with the employer
Schedule 1 para A.4.1	**Duty** Forthwith give written notice of extent, nature and location	If loss or damage affects the work or site materials
Schedule 1 para A.4.3	**Duty** Restore damaged work, replace or repair materials, remove debris and complete the Works	After any insurance inspection, with due diligence
Schedule 1 para A.4.4	**Duty** Authorise insurers to pay insurance monies to the employer	For contractor and all sub-contractors
Schedule 1 para B.2.1.1	**Power** Reasonably require the employer to produce documentary evidence of insurance	
Schedule 1 para B.2.1.2	**Power** Take out joint names insurance	If the employer defaults
Schedule 1 para B.2.2	**Power** Reasonably require the employer to produce a copy of the cover certificate relating to terrorism cover	
Schedule B.3.4	**Duty** Authorise insurers to pay insurance monies to the employer	For contractor and all sub-contractors
Schedule 1 para C.1	**Duty** Authorise insurers to pay insurance monies to the employer	For contractor and all sub-contractors
Schedule 1 para C.3.1.1	**Power** Reasonably require the employer to produce documentary evidence of insurance	Unless the employer is a local authority
Schedule 1 para C.3.1.2	**Power** Take out joint names insurance	Unless the employer is a local authority. If the employer defaults

Table 5.1 *Cont'd*

Clause	Power/duty	Precondition/comment
Schedule 1 para C.3.2	**Power** Reasonably require the employer to produce a copy of the cover certificate relating to terrorism cover	Where the employer is a local authority
Schedule 1 para C.4.1	**Duty** Forthwith give written notice of extent, nature and location	If loss or damage affects the work or site materials
Schedule 1 para C.4.3	**Duty** Authorise insurers to pay insurance monies to the employer	For contractor and all sub-contractors
Schedule 1 para C.4.4	**Power** Terminate the contractor's employment within 28 days of loss or damage	If just and equitable
Schedule 1 para C.4.4.1	**Power** Invoke the dispute resolution procedures	Within 7 days of receiving a termination notice
Schedule 1 para C.4.5.1	**Duty** Restore damaged work, replace or repair materials, remove debris and complete the Works	After any insurance inspection, with due diligence
Schedule 2 para 1	**Duty** Notify the architect of date on which named sub-contract is executed	
Schedule 2 para 2	**Duty** Immediately inform architect	If the contractor is unable to enter into the named sub-contractor in accordance with the particulars
Schedule 2 para 5.3	**Power** Make reasonable objection to entering into named sub-contract. **Duty** Enter into sub-contract using ICSub/NAM/A	Within 14 days of issue of instruction
Schedule 2 para 6	**Duty** Notify the architect of events likely to lead to termination of the named sub-contractor's employment	As soon as reasonably practicable

Table 5.1 *Cont'd*

Clause	Power/duty	Precondition/comment
Schedule 2 para 7	**Duty** Give the architect written particulars of the circumstances	If the named sub-contractor's employment terminates before completion of the sub-contract work
Schedule 2 para 10.2.1	**Duty** Take such reasonable action as necessary	To recover amounts from the named sub-contractor after termination under clause 7.4, 7.5 or 7.6 of ICSub/NAM/C
Schedule 2 para 10.2.2	**Duty** Account to the employer for amounts recovered	
Schedule 2 para 10.2.4	**Duty** Repay the employer additional amounts and pay or allow a sum equivalent to liquidated damages	If the contractor fails to comply with obligations to take reasonable action against the named sub-contractor for recovery

According to one line of authority, merely 'going slow' is not a breach of contract as such. It becomes a breach of contract only if there is ultimately a delay in completion. It is a question of fact whether the contractor is going ahead 'regularly and diligently', and this is clearly to be judged by the standards to be expected of the average competent and experienced contractor. The phrase 'regularly and diligently' has been defined in *West Faulkner* v. *London Borough of Newham* (1994), by the Court of Appeal:

> Although the contractor must proceed both regularly and diligently with the Works, and although each word imports into that obligation certain discrete concepts which would not otherwise inform it, there is a measure of overlap between them and it is thus unhelpful to seek to define two quite separate and distinct obligations.
>
> 'What particularly is supplied by the word "regularly" is not least a requirement to attend for work on a regular daily basis with sufficient in the way of men, materials and plant to have the physical capacity to progress the works substantially in accordance with the contractual obligations.'
>
> 'What in particular the word "diligently" contributes to the concept is the need to apply that physical capacity industriously and efficiently towards the same end.'
>
> 'Taken together the obligation upon the contractor is essentially to proceed continuously, industriously and efficiently with appropriate physical resources so as to progress the works steadily towards completion substantially in accordance with the contractual requirements as to time, sequence and quality of work.'

If the architect required the contractor to provide a programme, even though this is not a contract document, it is a standard against which the contractor's progress can be measured. However, failure by the contractor to meet the intermediate dates in the programme is unlikely to be enough, on its own, to evidence a failure to proceed regularly and diligently. The architect may require a contract programme by means of a clause in the bills etc. and the contractor is entitled to submit a programme showing completion date earlier than the contract completion date, and may complete the works earlier than the contract date if it wishes, whether or not the works are programmed. However, there is no obligation on the architect to provide information, etc. so as to enable the contractor to meet the earlier date: *Glenlion Construction Ltd* v. *The Guinness Trust* (1987) is decisive of this question. Clause 2.11.2 of the contract now incorporates this principle.

5.1.3 Workmanship and materials

Clause 2.2 deals with materials goods and workmanship. *Crown Estates Commissioners* v. *John Mowlem & Co Ltd* (1994) decided that the phrase at the beginning of this clause should be interpreted in a broad way with the effect that all quality and standards are to be to the architect's satisfaction. The clause goes on to say that if the quality and standards are not described in the contract documents nor stated to be for the architect's satisfaction, the standard must be 'appropriate for the Works'. This clause gives rise to severe difficulties. Although clause 1.10.1.1 makes clear that conclusivity with regard to the architect's satisfaction is dependent on whether a requirement for such satisfaction is expressly stated in the contract documents, no such limitation is put on the architect's satisfaction in the first part of this clause. Therefore reference to standards being appropriate to the Works if not stated to be a matter for the architect's satisfaction, in the second part of the clause, appears to be superfluous because it appears that they must always be to such satisfaction. A proper and strict reading of clauses 2.2.1 and 1.10.1.1 appears to be that quality and standards of all materials and workmanship must be to the architect's satisfaction, but it is in only those instances where the documents have expressly stated that such satisfaction is required that the final certificate will be conclusive that such satisfaction has been achieved.

The contractor is expected to show a reasonable degree of competence and to employ skilled tradesmen and others, although the architect has no power to direct how the work should be carried out.

The quality and quantity of work must be adequately defined in the contract documents, as is made plain by clause 4.1, the provisions of which are important in defining the contractor's obligations. In general, it may be said that any extra cost which results from faulty description in the contract documents falls on the employer. For example, if there is an inconsistency in the contract documents or an error of description, the architect must issue the appropriate instruction under clause 2.13.

Clause 2.2 should be studied carefully. Unlike SBC, IC makes no provision in the contract that the contractor is to provide materials, goods and workmanship 'so far as procurable', which would be a valuable protection. As it stands, the contractor's failure to supply would be a breach of contract. If it has become impossible to supply materials or goods of a particular kind, the contract may be frustrated if the materials are fundamental to the project: *Davis Contractors* v. *Fareham* (1956).

The other contractual references to materials and goods are in clauses 2.17 and 2.18, which are concerned solely with the transfer of ownership in materials and goods, and clause 4.12, which deals with payment for off-site materials and goods (see section 11.2.3). The statement in clauses 2.17 and 2.18 that 'such materials and goods shall become the property of the employer' is misleading, as this is not necessarily the case. If the contractor is not the legal owner of the goods, it cannot pass title to the employer. In practice, this means that the architect needs to take great care before including the value of unfixed goods or materials or off-site items in any interim certificate unless the contractor provides proof of ownership (e.g. a copy of the sale contract from the supplier). This is covered by clause 4.12 (see section 11.2.3). Many architects do not seem to appreciate the legal position.

5.1.4 Statutory obligations

Clause 2.1 places on the contractor an obligation to comply with all statutory obligations and, under clause 2.3, to pay all fees and charges in respect of the works which are legally recoverable from it (e.g. fees under the Building Regulations). Clause 2.15 also imposes on the contractor an obligation to give the architect immediate written notice on discovery of any divergence between the contract documents or an architect's instruction and the statutory requirements. The architect then has seven days in which to issue an instruction to deal with the divergence.

Contractually, clause 2.15.3 exempts the contractor from liability to the employer if the Works do not comply with statutory requirements, provided the work has been carried out in accordance with the contract documents or any architect's instruction where, for example, the contractor does not spot the divergence. The wording in clause 5.2 is 'if the Contractor or the Architect finds any divergence . . .', and, unless the contractor does so, it is under no obligation to notify the architect. This is clear and settled law: *London Borough of Merton* v. *Stanley Hugh Leach Ltd* (1985). However, whatever the position may be as between contractor and employer under the contract, the exempting provision cannot exempt the contractor from its duty to comply with the Building Regulations and other statutory obligations, liability under which may be absolute. But the wording is sufficiently wide to protect the contractor from any action by the employer, which may leave the architect in the firing line.

So far as fees and charges in connection with the works are concerned, these are to be added to the contract sum so that the contractor is reimbursed, although clause 2.3 establishes that the contract documents can require such fees and charges to be included in the tender sum. The reference to 'rates or taxes' is interesting and is designed to cover those (comparatively rare) occasions when site huts and so on are rateable.

5.1.5 Person-in-charge

Clause 3.2 requires the contractor to keep on site 'a competent person-in-charge'. This person must be there at all reasonable times, i.e. during normal working hours. This person is intended to be the contractor's full-time representative on site, but under this contract the appointment and replacement are not made subject to the architect's approval.

'Competent' can only mean what it says – i.e. having sufficient skill and knowledge – and the person-in-charge is the contractor's agent for the purpose of accepting the architect's written instructions, which are then deemed to have been issued to the contractor. If the person-in-charge or replacement is sensibly intended to be subject to approval by the architect, this is a matter which can be dealt with in the contract documents.

5.1.6 Levels and setting out

Although it is the architect's duty to determine any levels which may be required for the execution of the Works, and to provide the contractor with accurately dimensioned drawings to enable setting out to take place, clause 2.9 obliges the contractor to set out the work accurately in accordance with the architect's instructions. The contractor is made responsible for its own setting-out errors and must amend them at its own cost. Disputes may arise in this regard, and Figures 5.1 and 5.2 may be used by the contractor as appropriate. The sensible interpretation of the last sentence of clause 2.9 is that the architect may, with the employer's consent, instruct the contractor not to amend setting-out errors and make an appropriate adjustment to the contract sum, although exactly how this is to be assessed is not stated.

Figure 5.1
Contractor to architect if insufficient information on setting-out drawings

Dear

We are preparing to commence work on site on [*insert date*]. Our first task will be to set out the Works. An examination of the drawings you have supplied to us reveals that there is insufficient information for us to set out the Works accurately. We enclose a copy of your drawing number [*insert number*], on which we have indicated in red the positions where we need dimensions/levels [*delete as appropriate*].

We need this information by [*insert date*] in order to avoid delay and disruption to the Works.

Yours faithfully

Figure 5.2
Contractor to architect, requesting confirmation that setting out is correct

Dear

We refer to our letter of the [*insert date*] in which we notified you that the information on your drawings was insufficient to enable us to set out the Works accurately. You responded by telephone, asking us to set out the Works to the best of our ability based on the information provided. We have carried out your instructions, but necessarily we have had to make several assumptions.

We enclose our drawing S1 showing the dimensional location of the setting out pegs and we should be glad to receive your confirmation that the setting out is correct. If we do not receive such confirmation, in writing, by return of post, we will be unable to progress the Works and we shall be obliged to notify you of a delay to the Works and of disruption for which we will seek appropriate financial recompense.

Yours faithfully

5.2 *Other obligations*

5.2.1 Access to the Works and premises

A new clause (3.1) expressly provides that the architect or representatives should have access to the Works at all reasonable times. It is, of course, unnecessary for that right to be referred to expressly, because it is implied under the general law. But the architect also needs access to the workshops or other places of the contractor where work is being prepared for the contract. It is probable that the general law would not give the architect that right of access, and the new clause succinctly provides for this also.

5.2.2 Drawings, details and information

Where there is an information release schedule provided as stated in the sixth recital, the architect must provide the information stipulated at the times stipulated. The architect is excused from doing so if prevented by some act or default of the contractor or of any contractor's person (clause 2.10). It is difficult to envisage what such an act of prevention could encompass, because, under IC more than other contracts, the architect should be in a position to issue all information without any input from the contractor. The employer and the contractor may agree to vary the time of issue of the information.

Anecdotal evidence suggests that the provision of an information release schedule is not common. The principle of the provision of such a schedule contains a fundamental flaw. It is well established that the contractor is entitled to organise the work in whatever way it chooses and to decide, within reasonable limits, when information and details should be provided: *Wells* v. *Army & Navy Co-operative Society Ltd* (1902).

Not only must the contractor know at the time of tender that an information release schedule is to be provided, it must actually be provided so that the contractor can plan its work on the basis of information to be received. Clearly, this involves the architect being able to guess the way in which each separate tenderer intends to tackle the project. It also circumscribes the contractor's right to organise the work in its own way.

The alternative, an agreed schedule between architect and contractor shortly after the contract has been let, assumes a level and speed of agreement that experience teaches is unlikely.

If the information release schedule is not provided, clause 2.11 requires the architect to provide the contractor from time to time with further drawings, details and information enabling it to carry out and complete the Works by the date for completion, and failure to do this is a breach of contract for which the employer may in principle be liable in damages. But, under clause 2.11.3, the contractor is also under contractual duty to make a request for particular details or instructions if it knows when it needs the information and it has reasonable grounds for thinking that the architect does not know when the information is required. The contractor must advise the architect only to the extent that it is reasonably practicable to do so. There is no express requirement for the contractor to notify the architect in writing, but in practice the prudent contractor will continue the customary practice of notifying the architect in good time of all information required, and it will do it in writing if for no other reason than that there will then be a record of the notification.

Figure 5.3 is a suitable letter to the architect from the contractor. A suitably annotated and updated programme is likely to be a sufficient notification under this clause: *London Borough of Merton* v. *Stanley Hugh Leach Ltd* (1985).

Factors to be borne in mind include: the state the Works have reached, and whether the contractor can act on the information; the nature of the instruction or information; any other of the contractor's activities which may depend on the supply of information (e.g. pre-ordering of materials); and the time it may reasonably take for the architect to prepare the information: *Neodox Ltd* v. *Borough of Swinton & Pendlebury* (1958). On a strict reading of the clause it is likely that, where a contractor's progress is slow, the architect, who under clause 2.11.2 is to have 'regard' to such progress, is entitled to issue information to suit the slow progress. In practice, it may be difficult later to establish whether the information was issued in response to the slow progress or whether the slow progress is a result of late information. The prudent approach is for the architect to issue the information in such time as will allow the contractor to complete by the completion date irrespective of actual progress.

5.2.3 Compliance with architect's instructions

The architect's authority to issue instructions is limited to those instructions of which the contract empowers the issue. Under clause 3.8 the contractor has an obligation to obey all written instructions given by the architect which are authorised by the contract (see

Figure 5.3
Contractor to architect, requesting information

Dear

We hereby apply for the following information required by us for the carrying out and completion of the Works:

[*Description of information*] [*Date required*]

[*Alternative:*]

We hereby apply for the supply of [*state information required with supporting details showing why and when it is required*].

Yours faithfully

Figure 4.3). The contractor can require the architect to state in writing the contract clause under which the instruction is given (clause 3.10): see Figure 5.4. It also has a right to object to instructions falling within clause 5.1.2, which is concerned with alterations or obligations or restrictions imposed by the employer in the contract documents. Figure 4.13 is a suitable letter from the contractor.

The contractor must make 'reasonable objection in writing', but no guidance is given as to what is a reasonable objection. It is clearly not a reasonable objection that it is difficult for the contractor to comply, because the contractor is expected to overcome those difficulties, and questions of time and cost are dealt with under the contract. In fact, it seems that the instruction must make continued execution of the work almost impossible, e.g. by preventing deliveries to site by further restricting access.

The general sanction for non-compliance by the contractor is set out in clause 3.9. Under that provision, where the architect has served the contractor with a written notice requiring compliance and the contractor has not complied within seven days, the employer may engage others to carry out the work, and all the costs involved are to be deducted from the contract sum.

It is essential that, if the contractor does not comply with an instruction, the architect ensures that the remedy provided by the clause is put into operation. If the architect does not act, there is case law which suggests that the employer may then be taken to have waived rights under the clause, and, of course, the architect's notice of compliance may be regarded as becoming stale. (See also section 4.1.3.)

More effective sanctions are provided by clause 3.15, which deals with instructions following failure of work etc. This is a most useful provision (see Figure 4.18). It covers failure of work, materials or goods discovered during the carrying out of the Works. If, while the Works are being carried out, the contractor discovers that work, materials or goods are not in accordance with the contract, it must notify the architect in writing immediately. The clause says that the contractor must state in writing to the architect the action that the contractor will immediately take at no cost to the employer to establish that there is no similar failure in work already executed or materials or goods already supplied. This obligation extends to failures through the fault of any sub-contractor, named or otherwise. The clause goes on to deal with the architect's powers to issue instructions requiring opening up of work, etc. in three cases:

- Where the contractor's written statement has not been received within seven days of discovery of the failure; or

Figure 5.4
Contractor to architect, requiring him or her to specify the clause empowering an instruction

Dear

We have received today your instruction number [*insert number*], dated [*insert date*], requiring us to [*insert substance of instruction*].

We request you, in accordance with clause 3.10 of the conditions of contract, to specify in writing the provision which empowers the issue of the above instruction.

Yours faithfully

- If the architect is dissatisfied with the contractor's proposed action; or
- If the architect is unable to wait for the contractor's written proposals because of considerations of safety or statutory obligation (e.g. service of a dangerous structure notice or a prohibition notice under the Health and Safety at Work etc. Act 1974).

The architect's default powers are to issue appropriate instructions in writing, requiring the contractor, at its own cost, 'to open up for inspection any work covered up or to arrange for or carry out any test of any materials or goods . . . or any executed work to establish that there is no similar failure, including making good thereafter'. The contractor is bound to comply forthwith with such instructions, and, while clause 3.15.2 gives it the right to object, this is said to be 'without affecting his obligation' to comply. In other words, even if the contractor disagrees with the architect, it must carry out the instruction.

The contractor has 10 days from receipt of the architect's instruction under clause 3.15.1 to object to compliance, stating the reasons in writing. The architect must then consider the objections immediately and decide whether to withdraw or modify the instruction to meet the contractor's objection. If the architect takes neither action within seven days of receipt of the contractor's objection (and reasons), then any dispute or difference about whether the nature or extent of the opening up for inspection or testing instructed was reasonable in all the circumstances is referred to resolution under the relevant dispute resolution procedures. In the first instance the reference is likely to be to adjudication (see also section 4.1.4).

Figure 5.5 is a letter which the contractor might send to the architect after opening up work for inspection under clause 3.14.

5.2.4 Suspension of performance

Clause 4.11 provides the contractor with a valuable remedy against non-payment by the employer. If the employer does not pay the contractor in full by the final date for payment, the contractor may serve on the employer a written notice giving seven days in which to pay the amount owing. A copy must be sent to the architect. The notice must state the contractor's intention to suspend performance of its obligations (which will include the obligation to insure) and the grounds for so doing. The contractor may proceed to suspend its obligations if the employer does not pay within the seven days. The

Figure 5.5
Contractor to architect after work opened up for inspection

Dear

We confirm that you inspected [*describe work*] opened up for your inspection in accordance with your instruction number [*insert number*], dated [*insert date*], under clause 3.14 of the conditions of contract.

The inspection took place on the [*insert date*], and you found the materials, goods and work to be in accordance with the contract.

The cost of opening up and making good, therefore, is to be added to the contract sum and we will let you have details of our costs within the next few days. We will shortly send you details, particulars and estimate of the expected delay in completion of the Works beyond the completion date and an application for reimbursement under the appropriate clause of the contract.

Yours faithfully

contractor may continue the suspension until payment is made in full including VAT.

There are several points to note. The first is that the employer has not failed to pay an amount if a valid notice to withhold or deduct under clause 4.8.3 has been served. Although the clause is written in a way which appears to make the sending of a copy of the notice to the architect a condition precedent to suspension, the Housing Grants, Construction and Regeneration Act 1996 merely requires notice to be given to the other party. Therefore failure to give a copy to the architect will not prevent the contractor from lawfully suspending under the Act, although it may interfere with the contract mechanism for extension of time and loss and/or expense. On this point it is likely that the IC contract, like its predecessor IFC 98, fails to comply with the Act. The 1996 Act itself provides that, effectively, time is to be extended, however. The clause is without prejudice to the contractor's other rights and remedies, which means that it may choose to use its powers of termination instead or it may opt to terminate if suspension of obligations does not have the desired effect.

5.2.5 Other rights and obligations

Table 5.1 summarises the contractor's powers and duties generally. Other matters referred to in that table are dealt with in the appropriate chapters.

5.3 *Summary*

The contractor must:

- Carry out and complete the Works in accordance with the contract documents
- Proceed regularly and diligently with the Works so as to complete in due time
- Use workmanship and materials of an adequate standard
- Comply with relevant statutory obligations
- Obey architect's instructions as authorised by the contract
- Appoint and keep on site a competent person-in-charge.

The contractor may suspend performance if not paid.

CHAPTER SIX
THE EMPLOYER'S POWERS, DUTIES AND RIGHTS

6.1 *Express and implied powers and duties*

Like those of the contractor, some of the employer's powers and duties arise from the express provisions of IC itself. These are set out in Table 6.1.

Others are imposed by the general law by way of implied terms. These are provisions which the law writes into every building contract, and they apply so far as they are not excluded or modified by the express terms of the contract itself. In practice there are two important implied terms which are not affected by the contractual provisions.

6.1.1 Co-operation or non-interference

Under the general law it is an implied term in every building contract that the employer will do all that is reasonably necessary to bring about completion of the contract: *Luxor (Eastbourne) Ltd* v. *Cooper* (1941). Conversely, it is implied that the employer will not so act as to prevent the contractor from completing in the time and in the manner envisaged by the agreement: *Cory Ltd* v. *City of London Corporation* (1951). Breach of either of these implied terms which results in loss to the contractor will give rise to a claim for damages at common law.

If the employer – either personally or through the agency of the architect or that of anyone else for whom the employer is responsible in law – hinders or prevents the contractor from completing in due time, the employer is in breach of contract, and such conduct will also prevent the enforcement of the liquidated damages provision if any delay results.

The various cases put the duty in different ways, but in essence the position may be summarised as follows:

Table 6.1
Employer's powers and duties under IC and ICD (including powers and duties in the schedules)

Clause	Power/duty	Precondition/comment
2.5	**Power** Defer the giving of possession of the site or part	For a period not exceeding 6 weeks or the period in the contract particulars
2.6	**Power** Use or occupy the site or Works. **Duty** Notify the insurers under insurance option A, B or C and get confirmation that such use or occupation will not prejudice insurance	With the contractor's written consent before practical completion. Before such consent. The contractor may notify instead of the employer. Contractor's consent must not be unreasonably delayed or withheld
2.7	**Power** Require work not forming part of the contract to be carried out by the employer or employer's persons	If contract documents provide necessary information, the contractor must permit it. If contract documents do not provide information, contractor's consent is required, not to be unreasonably delayed or withheld
2.8.1	**Duty** Be custodian of the contract documents	Must be available for inspection by contractor at all reasonable times
2.8.3	**Duty** Not to divulge or use any of the contractor's rates or prices	Except for the purposes of the contract
2.9	**Power** Consent that errors in setting out are not to be amended	An appropriate deduction is to be made from the contract sum
2.10.1	**Power** Agree with the contractor to vary time of issue of information	In the information release schedule

Table 6.1 *Cont'd*

Clause	Power/duty	Precondition/comment
2.23.1	**Power** Give written notice to the contractor requiring payment of liquidated damages or that they will be withheld in respect of the period between completion date and practical completion	Not later than 5 days before the final date for payment of the final certificate provided that: • architect has issued certificate of non-completion; *and* • employer has notified contractor before the date of the final certificate that employer may deduct liquidated damages or require their payment
2.24	**Duty** Pay or repay any amounts deducted or recovered in respect of liquidated damages	If liquidated damages have been recovered, an extension of time has been given and the non-completion certificate has been cancelled
2.25	**Power** Take possession of part of the Works	If the employer so wishes and the contractor has given consent
2.30	**Power** Consent that defects are not to be made good	An appropriate deduction is to be made to the contract sum
2.32 (ICD only)	**Power** Reasonably require CDP documents or other information relating to the CDP as built	Does not affect the contractor's obligations under clause 3.18 regarding the health and safety file
2.33.1 (ICD only)	**Power** Copy and use the CDP documents and reproduce any designs in them	For any purpose relating to the Works except as a design for any extension of the Works
3.3	**Power** Appoint a clerk of works	To act as inspector under the direction of the architect
3.4.1	**Duty** Nominate a replacement architect or quantity surveyor **Duty** Nominate an acceptable replacement	If the architect or quantity surveyor ceases to hold the post. Must be done within 14 days of cessation. Except where the employer is a local authority and the replacement is an official, if the contractor for sufficient reason objects

Table 6.1 *Cont'd*

Clause	Power/duty	Precondition/comment
3.9	**Power** Employ and pay other persons to execute any necessary work in connection with an instruction	If the contractor has not complied with an instruction within 7 days of receiving a written notice
3.18.1	**Duty** Ensure that the planning supervisor carries out duties **Duty** Ensure that the principal contractor carries out duties	 If the contractor is not the principal contractor
3.19	**Duty** Immediately notify the contractor in writing of the name and address of new appointee	If the employer replaces the planning supervisor or the principal contractor
4.5	**Duty** Pay the advanced payment on the specified date	If the contract particulars so specify
4.8.2	**Duty** Give a written notice to the contractor specifying the amount to be paid	Not later than 5 days after the date of issue of the interim certificate
4.8.3	**Power** Give a written notice to the contractor specifying the amount to be withheld	Not later than 5 days before the final date for payment if the employer intends to withhold any payment
4.8.4	**Duty** Pay the amount in the clause 4.8.2 notice within 14 days of the date of issue of the certificate **Duty** Pay the amount on the certificate	If the architect issues a certificate If no notice issued
4.8.5	**Duty** Pay the contractor simple interest	If the employer fails to pay the amount due by the final date for payment
4.10	**Power** Have recourse to retention	Where the employer is not a local authority. For payment of any amount to which the employer is entitled under the contract provisions

Table 6.1 *Cont'd*		
Clause	**Power/duty**	**Precondition/comment**
4.14.2	**Duty** Give a written notice to the contractor specifying the amount to be paid	Not later than 5 days after the date of issue of the final certificate
4.14.3	**Power** Give a written notice to the contractor specifying the amount to be withheld	Not later than 5 days before the final date for payment if the employer intends to withhold any payment
4.14.4	**Duty** Pay any balance stated as due to the contractor within 28 days of the issue of the final certificate	If no clause 4.14.2 notice issued
4.14.5	**Duty** Pay the contractor simple interest	If the employer fails to pay the amount due by the final date for payment
6.4.2	**Power** Require the contractor to send documentary evidence of insurance to the architect. **Power** Require that relevant policies are sent to the architect	Must be exercised reasonably. At any time (but not unreasonably or vexatiously)
6.4.3	**Power** Insure and deduct premium amounts from monies due to the contractor or recover them as a debt	If the contractor fails to insure against personal injury or death or injury to property
6.5.2	**Power** Approve insurers	For insurance under clause 6.5.1
6.9.1	**Duty** Ensure that joint names policy *either*: • provides for recognition of each sub-contractor as an insured; *or* • includes a waiver of subrogation in regard to such sub-contractors.	Where insurance option B or C applies. In respect of loss or damage by specified perils to the Works or section. Continues to practical completion or termination if earlier

Table 6.1 *Cont'd*

Clause	Power/duty	Precondition/comment
6.9.3	**Duty** Also ensure that joint names policy *either*: • provides for recognition of each named sub-contractor as an insured; *or* • includes a waiver of subrogation in regard to such named sub-contractors	Where insurance option C applies. In respect of loss or damage by specified perils to the Works or section. Continues to practical completion or termination if earlier
6.10.1	**Duty** Inform the contractor	If insurers named in joint names policy notify the contractor that terrorism cover will no longer be available
6.10.2	**Duty** Give written notice to the contractor that *either* • the Works must be carried out; *or* • the contractor's employer will terminate on the date stated in the notice	If the employer receives notification from insurers that terrorism cover will cease
6.12	**Duty** Ensure compliance with Joint Fire Code of all employer's persons	
6.13.1	**Duty** Send copies of notice to contractor and architect	If breach of Joint Fire Code occurs and insurers notify the employer of remedial measures required
6.13.2	**Power** Employ and pay others to carry out the remedial work	If the contractor has not begun to carry out remedial measures within 7 days of receipt of the architect's clause 6.13.1.2 instruction
7.5	**Power** In a notice to the contractor identify purchaser or tenant and interest in the Works and require the contractor to enter into a warranty	Warranty must be executed within 14 days of receipt of notice using form CWa/P&T

Table 6.1 *Cont'd*

Clause	Power/duty	Precondition/comment
7.6	**Power** In a notice to the contractor identify funder and require the contractor to enter into a warranty	Warranty must be executed within 14 days of receipt of notice using form CWa/F
7.7	**Power** In a notice to the contractor identify the sub-contractor and require the sub-contractor to enter into a warranty with the purchaser, tenant or funder	If the contract particulars so provide. Warranty must be executed within 21 days of receipt of notice using forms CWa/P&T or CWa/F
7.8	**Power** In a notice to the contractor identify the sub-contractor and require the sub-contractor to enter into a warranty with the employer	If the contract particulars so provide. Warranty must be executed within 21 days of receipt of notice
8.4.2	**Power** Terminate the contractor's employment by written notice served by recorded, special or actual delivery	If the contractor has not ceased a specified default within 14 days of receiving a default notice
8.4.3	**Power** By written notice terminate the contractor's employment	Upon or within a reasonable time of the contractor repeating a default
8.5.1	**Power** At any time by notice terminate the contractor's employment	If the contractor is insolvent
8.5.3.3	**Power** Take reasonable measures to ensure that the site, Works and materials are protected	After insolvency
8.6	**Power** By written notice, terminate the contractor's employment	If the contractor commits a corrupt act

Table 6.1 *Cont'd*

Clause	Power/duty	Precondition/comment
8.7.1	**Power** Employ and pay others to carry out the Works. **Power** Take possession of the site and use all temporary building, etc.	If contractor's employment terminated under clauses 8.4, 8.5 or 8.6
8.8.1	**Duty** Forthwith notify the contractor in writing **Duty** Send a statement to the contractor setting out the total value of work properly executed and any other amounts due to the contractor under the contract and the aggregate amount of expenses and direct loss and/or damage caused to the employer	If, within 6 months from termination, the employer decides not to complete the Works. Within a reasonable time of such notification or, if no notice is given but the employer does not begin work during the 6 month period
8.10.2	**Duty** Immediately inform the contractor in writing	If employer makes any proposal, gives notice of meeting or becomes subject of proceedings, etc. relating to clause 8.1 matters
8.12.4	**Duty** Pay the contractor the amount properly due	Must be done within 28 days of submission by the contractor, after taking account of amounts previously paid, but without deducting retention
9.1	**Power** Agree to resolve a dispute by mediation	
9.3	**Power** By written notice jointly with the contractor to the arbitrator state that they wish the arbitration to be conducted in accordance with any amendments to the JCT 2005 CIMAR	
9.4.1	**Duty** Serve on the contractor a written notice	If the employer wishes a dispute to be resolved by arbitration

Table 6.1 *Cont'd*

Clause	Power/duty	Precondition/comment
9.4.3	**Power** Give a further arbitration notice to the contractor referring to any other dispute	After the arbitrator has been appointed. Rule 3.3 applies
Schedule 1 para A.1	**Power** Approve insurers	In regard to insurance against all risks for new buildings by the contractor
Schedule 1 para A.2	**Power** Take out insurance	If contractor fails to insure against all risks for new buildings
Schedule 1 para A.3	**Power** Inspect the policy and premium receipts for all risks insurance in joint names	At all reasonable times
Schedule 1 para A.4.5	**Duty** Pay insurance monies to the contractor by instalments under architect's certificates	Less only amounts in respect of professional fees
Schedule 1 para A.4.5	**Power** Retain amount properly incurred in respect of professional fees	From monies paid by insurers
Schedule 1 para A.5.2	**Power** Instruct the contractor not to renew terrorism cover under the joint names policy	Where the employer is a local authority, in lieu of adjusting the contract sum
Schedule 1 para B.1	**Duty** Take out and maintain joint names policy for all risks for new buildings	
Schedule 1 para B.2.1.1	**Duty** Produce documentary evidence of insurance and receipts for premium payments	As and when reasonably required by the contractor
Schedule 1 para B.2.2	**Duty** Produce copy of cover certificate regarding terrorism cover	Where the employer is a local authority as and when reasonably required by the contractor
Schedule 1 para C.1	**Duty** Take out and maintain joint names policy for existing structures	

Table 6.1 *Cont'd*

Clause	Power/duty	Precondition/comment
Schedule 1 para C.2	**Duty** Take out and maintain joint names policy for all risks for the Works	
Schedule 1 para C.3.1.1	**Duty** Produce documentary evidence of insurance and receipts for premium payments	As and when reasonably required by the contractor
Schedule 1 para C.3.2	**Duty** Produce copy of cover certificate regarding terrorism cover	Where the employer is a local authority as and when reasonably required by the contractor
Schedule 1 para C.4.4	**Power** Terminate the contractor's employment within 28 days of loss or damage	If just and equitable
Schedule 1 para C.4.4.1	**Power** Invoke the dispute resolution procedures	Within 7 days of receiving a termination notice
Schedule 2 para 3	**Power** Have omitted work carried out by a directly employed person	If the instruction is under paragraph 2.2
Schedule 2 para 8	**Power** Have omitted work carried out by a directly employed person	If the instruction is under paragraph 7.3
Schedule 2 para 10.2.3	**Power** Agree to indemnify the contractor against legal costs	Incurred in relation to taking action for recovery of monies from the named sub-contractor after termination

- The employer and employer's persons must do all things necessary to enable the contractor to carry out and complete the Works expeditiously and in accordance with the contract
- Neither the employer nor the employer's persons will in any way hinder or prevent the contractor from carrying out and completing the Works expeditiously and in accordance with the contract.

The scope of these implied obligations is very broad, and in recent years more and more claims for breach of them have been before arbitrators or the courts. The employer must not, for example, attempt to dictate how the architect is to exercise discretion, nor must the employer attempt to give direct orders to the contractor. So important is the architect's independent role under the contract that, for example, it has been held that the architect is entitled to terminate an engagement if the employer interferes with the granting of extensions of time under the contract: *Argyropoulos & Pappa* v. *Chain Compania Naviera SA* (1990). Similarly, the employer must see that the site is available for the contractor and that access to it is unimpeded by employer's persons: *Rapid Building Group Ltd* v. *Ealing Family Housing Association Ltd* (1984). This is especially important in works to existing structures or tenanted buildings.

Some potential acts of hindrance or prevention by the employer are covered by express clauses in the contract, but there are several grey areas.

6.2 *Rights*

6.2.1 General

Although the contract is between the employer and the contractor – who are the only parties to it – an analysis of the contract clauses shows that the employer has few express rights of any substance.

The employer's major right is, of course, to receive the completed Works in due time, properly completed in accordance with the contract documents. But the other rights are of importance as the contract proceeds.

6.2.2 Deferment of possession of the site

Clause 2.5 confers a right which the employer would not otherwise possess, namely the right to defer giving possession of the site or, if the Works are in sections, any relevant part to the contractor for a period up to six weeks (assuming that the contract particulars state that this provision is to apply). This is an important right in practice because, under the general law, failure to give the contractor sufficient possession to enable the Works to progress is a serious breach of contract. Possession of the site has been held to mean possession of the whole of the site: *Whittal Builders* v. *Chester-Le-Street District Council* (1987).

This power will be especially helpful in renovation works, for example, although the normal intention must be that it is to be exercised sparingly, because, if one has reached the contract stage, it is to be assumed that sufficient possession will be given to the contractor on the due date. Certainly, in the absence of such a provision, there would be no power to defer or postpone giving possession, and it would be necessary for the employer and the contractor to reach a separate agreement. Clauses 2.20.6 and 4.18.5 deal with the contractor's entitlement to extension of time and loss and/or expense in such circumstances.

6.2.3 Deduction/repayment of liquidated damages

If the contractor is late in completing the Works, then – provided the architect has issued a certificate of non-completion under clause 2.22 – the employer is entitled to recover liquidated damages at the rate specified in the contract particulars. This is usually done by deduction from sums due to the contractor (e.g. under interim certificates), but if (unusually) no sums are due to the contractor, then the employer must sue for them as a debt.

There are two further preconditions to deduction or recovery. Clause 2.23 makes it plain that payment of liquidated damages is not obligatory. The mere fact of late completion and the issue of the clause 2.22 certificate are not sufficient. The employer must give notice to the contractor, not later than the date of the final certificate requiring payment or stating that payment may be withheld before the issue of the final certificate, of the intention to exercise discretion to claim or deduct liquidated damages, because the employer is required to indicate to the contractor whether the liquidated damages are to be allowed or to be paid and the amount it is intended to deduct or claim: *J.J. Finnegan Ltd* v. *Community Housing Association Ltd* (1995). In addition, the employer must give an additional notice under clause 4.8.3 or 4.14.3 not later than five days before the final date for payment under the final certificate. Strangely, the employer is given no express discretion to deduct at a rate lower than the rate in the contract particulars, although it is doubtful that the contractor would object.

Figure 6.1 is a suitable letter from the employer to the contractor requiring liquidated damages. As liquidated damages have been held by the Court of Appeal to be exhaustive of the employer's rights for the breach of late completion, it is clear that an employer who has not exercised the right to deduct, or has failed to give the requisite notice to the contractor, cannot recover ordinary unliquidated damages for late completion: *Temloc Ltd* v. *Errill Properties Ltd* (1987).

If for some reason the liquidated damages clause fails – for example, if the employer causes delay to completion for which no extension of time is grantable under the contract or for some technical reason – the employer can then recover general damages by way of arbitration or litigation, subject to proof of actual loss. It is likely that the amount so recoverable will be held to be limited to the amount of the failed liquidated damages clause. Anything else would be an open invitation to take steps to invalidate the liquidated damages clause whenever it seems likely that the actual loss will greatly exceed the liquidated damages inserted in the contract.

The employer must repay any liquidated damages which have been recovered if the architect makes a further extension of time and thereby cancels the non-completion certificate (clause 2.24). Undoubtedly, contractors will argue that any damages so repaid should attract interest, and many ingenious arguments are likely to be advanced to support the contention, notably with reference to a decision of the High Court of Northern Ireland: *Department of the Environment* v. *Farrans (Construction)* (1982). Specialist practitioners are generally agreed that, even if correct on its facts, that decision is not good law generally – and it certainly has no application to IC. If interest was payable, this would be stated in clause 2.24. Should the contractor advance such an argument, Figure 6.2 is the sort of reply which the employer might send. The architect has nothing to do with the deduction or enforcement of liquidated damages.

If the architect purports to deduct liquidated damages on a certificate, Figure 6.3 is the sort of letter a contractor might send. Figure 6.4 is a letter from a contractor to an employer who has wrongfully deducted liquidated damages.

Figure 6.1
Employer to contractor, initial notice regarding recovery of liquidated damages

SPECIAL DELIVERY

Dear

The architect having issued the certificate of non-completion under clause 2.22 of the conditions of contract, certifying that you have failed to complete the Works by [*insert date as certified*], in accordance with clause 2.23.2, take this as formal notice that I may require payment of, or may withhold or deduct, liquidated damages at the rate of £[*insert amount*] for every [*insert 'day', 'week', etc*] during which the Works remain/ have remained [*delete as appropriate*] uncompleted.

Yours faithfully

Figure 6.2
Employer to contractor if contractor claims interest on liquidated damages repaid

Dear

Thank you for your letter of the [*insert date*] in which you claim that you are entitled to interest on the liquidated damages repaid / which will be repaid [*delete as appropriate*] to you as a result of the further extension of time dated [*insert date*].

I formally deny that you are entitled to the interest claimed or to any interest at all. The contract makes no provision for the payment of interest in those circumstances, and interest is not recoverable under the general law, because the liquidated damages now repaid was originally deducted in accordance with the provisions of the contract.

Yours faithfully

Figure 6.3
Contractor to architect who wrongfully deducts liquidated damages

Dear

We have received a copy of your certificate number [*insert number*] in which you purport to deduct the sum of £x as liquidated damages.

This deduction is improper, because under clause 2.23 it is for the employer to deduct liquidated damages, and the contract does not give the architect power to do so. In any event, you have not issued a certificate of non-completion under clause 2.22, and so any purported deduction by the employer would also be wrongful. Will you please issue an amended certificate in proper form?

Yours faithfully

Figure 6.4
Contractor to employer where liquidated damages wrongfully deducted

Dear

Thank you for your cheque in the sum of £ [*insert amount*] in respect of the payments due to us under certificate number [*insert number*]. Your cheque falls short of the sum certified as due to us by £ [*insert amount*], and we note that this deduction is purportedly made in respect of liquidated and ascertained damages of £ [*insert amount*] for [*insert number*] weeks.

We are advised that this deduction is a breach of contract on your part because [*state reasons, e.g., the architect has not certified that there is a delay in completion*], and we look forward to receiving your cheque for £ [*insert amount*], being the balance owing to us.

Failing receipt of that sum within 7 days from today's date there will be a dispute between us, and we shall refer the matter to adjudication for the recovery of this amount together with interest and costs. We also reserve the right to terminate our employment under clause 8.9 of the contract or to suspend performance of our obligations under clause 4.11.

Yours faithfully

Copy: Architect

6.2.4 Employment of direct contractors

Clause 2.7 is an important provision, because under it the employer has the right – if the contract documents so provide – to carry out or have carried out by others 'work not forming part of this Contract'. The contractor is obliged to permit this work to be executed while the contract Works are in progress, provided that reference is made to it in the contract documents. If no such reference is made, the employer can still have such work carried out with the consent of the contractor, which must not be unreasonably withheld.

This is an important right, which the employer certainly would not have at common law, because, as explained in Chapter 9, the contractor is in principle entitled to exclusive possession of the site during the currency of the contract. But the limitations of the provision must be noted. It applies only to work 'not forming part of this Contract', and cannot be used to take work away from the contractor.

This is subject to the limited provision in paragraphs 2 and 8 of schedule 2 in relation to the work of named sub-contractors where the contractor has, for good reason, been unable to enter into a sub-contract with the person named and the architect has either instructed its omission or omitted the work from the contract documents and substituted a provisional sum. In either event, the contractor is entitled to an extension of time and money.

The fundamental principle is that a clause of this sort cannot in general be used to omit contract work to give to others, because the contractor has agreed to do a certain quantity of work and has a right to do it. But a provision in terms of clause 2.7 – sometimes called an 'Epstein clause' – does enable the employer to carry out work while the contract is in progress – e.g. through the employer's own direct works department – or to have similar work carried out by others.

The words 'not forming part of this Contract' mean exactly what they say. It is not work which the employer can require the contractor to do, and it is important to get this matter clear because of the implications in time and cost. Under clauses 2.20.6 and 4.18.5 the contractor is entitled both to extension of time for any consequent delay and also to a loss and/or expense claim. Statutory undertakers – water suppliers, gas suppliers and the like – may fall under this clause if they are in a direct contract with the employer, but not where they are carrying out work 'in pursuance of (their) statutory obligations'.

The employer's power directly to employ and pay others to execute the work (clause 3.9) where the contractor has failed to comply with the architect's notice requiring compliance with an instruction does not fall under clause 2.7.

The use of clause 2.7 may be one method of avoiding some of the difficulties thrown up by clause 3.7 and schedule 2 with regard to named sub-contractors because, by careful forethought, specialist work can be made the subject of direct employment by the employer: in that case the employer is solely responsible for the acts or defaults of such direct contractors and has no action against the contractor, whether for insurance purposes or otherwise.

6.2.5 Rights as to insurance

The complex insurance clauses of IC were discussed in Chapter 3, but the employer's rights relating to insurance should be noted.

Clause 6.4.3 and paragraph A.2 of schedule 1 give the employer a default power to insure or maintain policies in force where the contractor fails to do so and to recoup the cost out of the monies due or to become due to the contractor.

Paragraph C.4.4 of schedule 1 – Special power of termination (see section 12.1.6): the power to terminate the contractor's employment can be exercised only if it is just and equitable so to do, a matter which an adjudicator may have some difficulty in deciding.

The employer's other contractual rights are summarised in Table 6.1, where they are described as 'powers'. They are discussed in the appropriate chapters.

6.3 *Duties*

6.3.1 General

The essence of a duty is that it must be carried out. It is not permissive but mandatory, and breach of a duty imposed by the contract will render the employer liable in damages to the contractor for any proven loss.

Not every breach of a contractual duty will entitle the contractor to treat the contract as being at an end; only breach of a provision which goes to the root or basis of the contract will do that, for example permanently stopping the contractor from entering the site. But a breach of any contractual duty will always, in theory, entitle the contractor to at least nominal damages, although in many cases any loss will be difficult if not impossible to quantify.

6.3.2 Payment

From the contractor's point of view the most fundamental duty of the employer is to make payment in accordance with the terms of the contract. However, although steady payment of certificates is essential from the contractor's point of view, the general law does not generally regard failure to pay, or to pay on time, as a major breach of contract. Repeated failure by the employer to pay the amounts certified by the architect would probably constitute a repudiatory breach of contract: *D.R. Bradley (Cable Jointing) Ltd* v. *Jefco Mechanical Services* (1988). It has more recently been held that just one failure to pay, coupled with a threat to pay no more until the Works were completed, will also amount to a repudiatory breach: *C.J. Elvin Building Services Ltd* v. *Noble* (2003).

IC is quite specific about payment (see Chapter 11). The basic provisions are contained in clauses 4.8 and 4.9 (interim payments) and clause 4.14 (final payment). It is established that under the general law the employer may withhold payment of certified sums in closely defined circumstances: *C.M. Pillings & Co Ltd* v. *Kent Investments Ltd* (1986); *R.M. Douglas Construction Ltd* v. *Bass Leisure Ltd* (1991). Following the coming into force of the Housing Grants, Construction and Regeneration Act 1996, express provisions were inserted in the contract. If the employer feels justified in withholding payment certified in an interim certificate or in the final certificate, notice is required indicating the grounds and the amount to be withheld or deducted under each ground (clauses 4.8.3 and 4.14.3).

If the employer fails to carry out the notice procedure correctly before withholding payment, the contractor is entitled to exercise a contractual right to suspend work until payment in full is made (clause 4.11) or to serve a default notice before terminating its employment under clause 8.9. In addition, the employer is liable to pay interest on amounts not paid in due time (clauses 4.8.5 and 4.14.5). These procedures are set out in detail in Chapter 11.

Once the architect has issued an interim certificate (clauses 4.8 and 4.9), the employer is given a period before the certificate must be paid. Payment must be made by the final date for payment, which is 14 days after the date of the certificate. If the certificate is sent to the employer by post, then effectively there are 13 days for paying the amount due – assuming, of course, that the certificate is delivered the next day. Within five days of the date of issue of a certificate, clause 4.8.2 requires the employer to give written notice to the contractor. The notice must state the amount of payment to be made in respect of the amount stated as being due in the certificate. This procedure seems to provide the employer with a mechanism to abate the sum certified, although it is not certain that this effect was intended.

The effect is reduced, however, because the second notice which the employer may give no later than five days before the final date for payment refers to the amount to be withheld from the amount due or from the earlier notified amount unless that is deemed to be the amount due. Where payment follows an architect's certificate, the amount due is the amount in the certificate: *Rupert Morgan Building Services (LLC) Ltd* v. *David Jervis and Harriett Jervis* (2004). Note that the contractor has powers of suspension (see section 5.2.4) if the employer does not pay in full by the final date for payment, subject only to the clause 4.8.3 or clause 4.14.3 notice. The provision leaves somewhat open whether payment in full refers to the certified sum less any properly notified deduction or withholding of payment, but including interest.

Payment by cheque is probably good payment, although some contractors have been known to argue to the contrary under other forms of contract. It is not permitted for the employer to say, for example, that computer arrangements do not fit in with the scheme of the certificates. If this is so, then the payment period should have been amended before the contract was let.

For the final payment (clause 4.14), the period between the date of issue and the date of final payment is 28 days. By the terms of the contract, the employer is entitled to deduct from both interim and final certificates certain specific and ascertained sums, notably liquidated and ascertained damages under clause 2.23.

The contract also contains other express provisions empowering deductions. These are clause 6.4.3 – insurance premiums paid on contractor's default; paragraph A.2 of schedule 1 – insurance premiums similarly paid; clause 8.7.4 – payments on termination of employment.

Despite the clear provisions of the contract, contractors sometimes find it difficult to secure payment. Sometimes the employer pays less than the amount certified by deducting sums to which there is no entitlement. Figure 6.5 is a pro forma for use by the contractor in such circumstances.

6.3.3 *Retention*

The employer has certain rights in the retention percentage. The retention monies are a trust fund except where the employer is a local authority, as is manifest by the wording of clause 4.10, which states that 'Where the Employer is not a local authority the

Figure 6.5
Contractor to employer where payment not made in full

SPECIAL DELIVERY

Dear

We have received your cheque for £ [*insert amount*], which is [*insert amount*] less than the sum certified as due to us by the architect in certificate number [*insert number*]. In your accompanying letter you state that you are withholding the balance because [*state reasons given by employer*].

The contract does not permit you to make this deduction, and therefore, your action in doing so amounts to a breach of contract. In addition, we remind you that the contract makes provision for interest on late payment at a high rate under clause 4.8.5, and that, under clause 4.11, we have the right to suspend performance of all our obligations (which of course includes the obligation to insure) on 7 days' notice. Whether or not we avail ourselves of those remedies, we have the right to issue a default notice prior to termination of our employment under clause 8.9.

If we do not receive the sum of £ [*insert amount*], being the balance owing to us, within 7 days from the date of this letter, we must regretfully invoke our contractual right under clause 4.11 to suspend performance of our obligations until payment is made in full.

Yours faithfully

Copy: Architect

Employer's interest in the [retention percentage] shall be fiduciary as trustee for the Contractor'. The contractor's interest in the retention percentage is subject only to the employer's right to take from it in respect of an amount which the contract allows the employer to withhold or to deduct from a sum which is or which may become due to the contractor. This right protects the employer if a question arises about the propriety of taking money from retention. The clause provides that the employer has no obligation to invest, which sits awkwardly with the usual duties of a trustee as noted below.

The deductions which the employer may make from the percentage retained are those referred to in section 6.3.2. As the retention money is trust money, it is not the employer's property – hence the need to confer on the employer the contractual right to deduct from it. This apart, the employer has no legal or other interest in the retention, and can be ordered by the court – on the contractor's application – to set the percentage withheld aside in a separate bank account. Although there is no express contractual provision requiring the employer to put the monies aside in a separate bank account, it is suggested that this can be required, and Figure 6.6 may be utilised: *Rayack Construction Ltd* v. *Lampeter Meat Co Ltd* (1979). Indeed, it is likely that the employer has a duty to put the money in a special account whether or not requested to do so, and the courts readily imply a term to that effect even where the term existed and was struck out by the parties: *Wates Construction (London) Ltd* v. *Franthom Property Ltd* (1991). Nothing is said about whether the contractor is entitled to interest on the percentage withheld, and it has been convincingly argued about a similarly worded clause that the contractor is entitled to interest because the employer is in the position of a trustee of the money. The statement that the employer has no obligation to invest may be contrary to the Trustees Acts.

6.3.4 Other duties

These are summarised in Table 6.1, and are commented on as necessary in appropriate chapters.

6.4 Summary

The employer is under a duty in common law to do all that is reasonably necessary to bring about completion of the contract. Under the contract, the employer has a right to:

Figure 6.6
Contractor to employer, requesting that retention money be placed in a separate bank account – private employers only

Dear

As you are aware, clause 4.10 of the conditions of contract provides that the retention money is trust money to which we are beneficially entitled and for which you are trustee for us.

We formally request you to place all the retention money included in the next and all future certificates in a separate bank account set up for that purpose and designated as a trust account in our favour as required by law. Please inform us of the name and address of the bank, the designation of the account, and its number.

Yours faithfully

- Defer giving possession of the site to the contractor for a maximum period of six weeks
- Recover liquidated damages for late completion
- Employ direct contractors to carry out work 'not forming part of the contract'
- Insure in default of the contractor so doing.

The employer must as a duty:

- Pay the contractor in accordance with the contract terms
- Observe all the contract provisions.

CHAPTER SEVEN
THE CLERK OF WORKS

7.1 *Appointment*

The appointment of a clerk of works is a matter for the employer, acting on the advice of the architect. One should certainly be appointed where the nature of the work demands constant or frequent inspections, and it is normal practice for the clerk of works to be appointed directly by the employer. Some organisations have their own clerks of works on their permanent staff. A few firms of architects employ clerks of works directly, but in light of the case law it is inadvisable for the clerk of works to be the employee of the architect.

Whether a clerk of works is necessary on a full-time or part-time basis – or at all – is a matter about which the architect must decide and advise the employer at tender stage. Where the architect's contract with the employer incorporates the Conditions of Engagement CE/99 model letter and services supplement this makes it clear that the architect is required to make visits to site, but this does not involve frequent or constant inspections. If that service is needed, then clause 3.10 of the conditions included in CE/99 provides that the client will appoint full-time or part-time site inspectors (defined as 'Clerks of Works or others').

7.2 *Duties*

The duties of the clerk of works are set out in the contract, clause 3.3:

> 'The Employer shall be entitled to appoint a clerk of works whose duty shall be to act solely as inspector on behalf of the Employer under the directions of the Architect.'

The contract makes no other reference to the clerk of works – not even to give power to issue directions (as in SBC). The only function of the clerk of works is to inspect, i.e. to examine closely for faults or errors. Although the clerk of works is often referred to as the architect's eyes and ears on site, the contract makes it clear that these functions are carried out as inspector on behalf of the employer and not the architect.

Because the clerk of works may only inspect, the subject of any directions from the architect must, presumably, be how and where such inspection should take place, and to what, if anything, particular attention should be paid. Although it is traditional for the architect to give directions to the clerk of works in a fairly informal manner, written confirmation of any directions which are other than routine should always be given. It is generally accepted, by architects, contractors and clerks of works alike, that the clerk of works will do more than simply inspect. For example, the architect will often ask for measurements or levels, and the contractor may rely on the clerk of works to solve minor problems on site. It must be understood, however, that the contract recognises none of this.

If the clerk of works issues any instructions or directions to the contractor, they should be ignored. If a contractor complies with an instruction given by the clerk of works and is involved in extra cost, it cannot claim reimbursement. Problems between contractors and clerks of works are not uncommon, because some clerks of works appear unaware of the limited nature of their duties and rights. The clerk of works may, of course, make some useful suggestions to both architect and contractor, particularly providing a valuable service by spotting mistakes on drawings. A competent and experienced clerk of works can generally assist in the smooth running of the contract.

It is not at all certain that the common practice of many local authorities of having specialist clerks of works is permitted under the contract. The contract clearly provides for one clerk of works. There are some sites where visits by specialist trades clerks of works and a chief clerk of works, as well as the regular clerk of works, were commonplace. A contractor who has tendered on the basis of a contract provision for one clerk of works may well be able to deny entrance to the site to all but one clerk of works or, alternatively, to found a claim for disruption and additional payment.

From the outset, however, it is essential that everyone fully understands the clerk of works' role, and this is a matter best dealt with at the first site meeting and clearly recorded in the minutes. The Institute of Clerks of Works considers that the contractual provisions are inadequate, and sometimes clause 3.3 may be amended. The provision should be left exactly as it stands or else made much more comprehensive, and any amendment should be undertaken only after obtaining advice from a specialist contract expert.

Irregularities do occur. It is, for example, a common practice for the clerk of works to make marks in chalk or wax crayon to indicate defects. It has been known for a clerk of works to deliberately deface unsatisfactory materials and goods to ensure that they are not incorporated or, if already incorporated, to ensure that they are replaced. Clerks of works are not entitled to take this kind of action. If they do, employers could find themselves facing a large bill from the contractor. Clerks of works would probably argue that they cannot possibly do harm to something which is already defective. They should remember that defective materials are not, by definition, the property of the employer, and that contractors may – and usually do – find another use for them elsewhere. Because they are contractors' materials, they are entitled to charge (perhaps, on occasion, a clerk of works personally) if they are wilfully damaged in this way.

Clerks of works should not issue what are usually termed 'snagging lists'. The clerk of works is on site for the benefit of the employer. Naturally, the clerk of works must point out major defects and draw attention to bad materials and so on. Similarly, the contractor's attention must be drawn to the host of minor defects which are often present. But being more specific has two dangers. One danger from the employer's point of view is that the contractor may consider that, if it rectifies everything on a clerk of works' list, it has fulfilled its contractual obligations. That is not what the contract says, but disputes, however misguided, may follow. The second danger is that by acting in this way the clerk of works is doing the job that properly belongs to the person in charge at the employer's expense.

Figures 7.1 and 7.2 may be used by a contractor as a basis for dealing with two common problem situations.

7.3 *Responsibility*

Clerks of works have a responsibility to carry out their duties in a competent manner. They are expected to show the same degree of skill as would be shown by the average clerk of works. If a clerk of works holds himself or herself out as being specially qualified in some branch of the industry, greater skill will be expected in that respect.

The architect is primarily responsible for inspection of the Works. The question that frequently arises is the extent to which such responsibility might be reduced or even extinguished by the employment of a clerk of works. It is now settled that, where the clerk of works is employed by the employer, the employer is

Figure 7.1
Contractor to architect where clerk of works issues instructions or directions to contractor

Dear

The clerk of works has directed us to [*specify*],

[*or:*]

The clerk of works has issued what is described as an instruction requiring us to [*specify*],

[*then:*]

a copy of which is enclosed. As you are aware, such a direction/instruction has no contractual effect, and we are not acting upon it. We accept that the clerk of works is bound to point out to us any defects in workmanship or materials, but we are only contractually obliged to act on instructions which are issued by you.

However, we are anxious to avoid any misunderstandings, and in this spirit we suggest that you ask the clerk of works to issue no further directions or instructions. Instead, perhaps you can instruct that any alleged difficulties should be referred to you by telephone so that, at your discretion, a proper architect's instruction can be issued.

Yours faithfully

Figure 7.2
Contractor to architect if clerk of works defaces work or materials

Dear

We regret to inform you that the clerk of works has today defaced [*specify*], presumably on the basis that the materials are defective. On our enquiry, the clerk of works informed us that the action was taken in order to bring the matter to our attention and to ensure that the materials were removed forthwith.

We object to this practice on two grounds:

1. The work or materials so marked may not in fact be defective under the terms of the contract, and we will be involved in extra work and the employer in extra costs in such circumstances.

2. The defaced work or materials have not been paid for, and they will be our property if removed. Defacement by the clerk of works prevents us reusing such materials in a less stringent situation if, indeed, they are defective.

We shall be glad if you will instruct the clerk of works to desist from this sort of action. If the practice continues, we will seek reimbursement from the employer for the costs involved.

Yours faithfully

vicariously responsible for the clerk of works' actions. This is so even though the clerk of works is acting under the directions of the architect. This has an important practical consequence should the work be supervised negligently and a claim be made by the employer against the architect. Although the employment of a clerk of works does not reduce the architect's responsibility to use reasonable skill and care to see that the Works conform with the design, if the clerk of works is also guilty of negligent inspection this would amount to contributory negligence for which the employer is responsible, and thus reduce the extent of the negligent architect's financial liability to the employer: *Kensington & Chelsea & Westminster Area Health Authority* v. *Wettern Composites & Other* (1984).

7.4 *Summary*

- The clerk of works should be appointed by the employer
- Necessary if frequent or constant inspections are required
- The appointment should be immediately after the successful tender has been accepted
- The clerk of works should be thoroughly briefed by the architect
- The clerk of works is purely an inspector, who cannot give instructions or directions to the contractor
- The clerk of works is not entitled to deface or put any mark on the contractor's work or materials
- The clerk of works must carry out his or her duties competently
- The employer is vicariously responsible for any negligence by the clerk of works.

CHAPTER EIGHT
SUB-CONTRACTORS AND SUPPLIERS

8.1 *General*

This chapter deals with the contract provisions for sub-contractors, suppliers, statutory authorities, and persons engaged by the employer to carry out work not forming part of the contract.

Provision is made for the employer to name sub-contractors to carry out work priced in the contract documents or included as a provisional sum (see section 8.2). There are no similar provisions for suppliers.

8.2 *Sub-contractors*

8.2.1 Assignment and sub-contracting

IC contains the usual restriction (clause 7.1) on the assignment of the contract by either party without the written consent of the other. Without this express term, it would be possible for either party to assign the benefits of the contract to another. For example, the contractor might wish to assign to a third party the benefit of receiving payments under interim certificates, in return for which the third party would give the contractor financial advances to enable it to carry out the work; this is a well-known procedure. Or the employer might wish to sell the building before the issue of the final certificate, or when the structure is only partly completed. Even without the express term, it is a matter of general law that the burden of a contract cannot be assigned without the consent of the other party. For example, the contractor cannot transfer to another the burden of carrying out the work, and the employer cannot pass to another the obligation to pay the contract sum. This type of clause has been held to be effective in preventing assignment of the benefits of a contract or the right to damages for breach of contract: *Linden Gardens* v. *Lenesta Sludge Disposals* (1993) and *St Martin's Property Corporation* v. *Sir Robert McAlpine & Sons* (1993).

There are very real difficulties for both parties in this clause, because the party withholding consent is not required to be reasonable in doing so. The architect should warn the employer and seek instructions on amending the clause if the employer is likely to want to assign the benefit before the time for issuing the final certificate. If the employer wants to amend the clause, it is a matter for specialist advice.

In contrast, if the contractor wishes to sub-let any part of the Works, the architect must not withhold consent unreasonably. The contractor does not have to inform the architect of the name of the sub-contractor, but the architect would probably be reasonable in withholding consent if the contractor did not do so (clause 3.5). A suitable letter for the contractor to write is shown in Figure 8.1. It is probably advisable for the architect to make the consent conditional upon the contractor acknowledging liability for its sub-contractors: *Scott Lithgow* v. *Secretary of State for Defence* (1989) (Figure 8.2).

The employer has no contractual relationship with the sub-contractor, and the contractor must bear liability for any defects in the sub-contractor's work. The employer will look to the contractor for redress in such circumstances. The contractor must, in turn, look to the sub-contractor. Any dispute, difficulty or difference arising between contractor and sub-contractor is a matter solely for the parties involved. The architect should beware of being drawn into such disputes. Despite that, clause 3.6 sets out certain provisions which the contractor must ensure are included in any sub-contract. It would not be unreasonable if the architect made consent to sub-letting subject to the contractor providing documentary evidence that these provisions are to be included, because contractors commonly sub-contract on their own terms. There is a proviso (clause 3.6.1) that the sub-contract must provide for the employment of any sub-contractor to terminate immediately on the termination of the contractor's employment for whatever reason. This is a perfectly sensible provision to prevent the situation arising in which the contractor's employment is terminated and the sub-contractor is able to sue the contractor on the ground of prevention from carrying out the sub-contract.

Clause 3.6.2 relates to the ownership of unfixed materials on site. The intention is to prevent the sub-contractor from reclaiming goods delivered to site which have already been paid for by the employer under the architect's certificate. Clause 3.6.2.1 requires the contractor's consent to the sub-contractor removing such materials from site, and, by clause 3.6.3, this is made subject to main contract clause 2.17, which requires the architect's consent (see section 4.2.1).

Figure 8.1
Contractor to architect, requesting permission to sub-let

Dear

We propose to sub-let portions of the Works as indicated below because [*state reasons, which may be, for example, because the required skills are not present within the contractor's organisation*]. We should be pleased to receive your consent in accordance with clause 3.5 of the contract.

Portion to be sub-let　　Name of proposed sub-contractor

[*Complete as appropriate*]

Yours faithfully

Figure 8.2
Architect to contractor, consenting to the use of sub-contractors

Dear

Thank you for your letter of the [*insert date*] requesting my consent to sub-letting of the following portions of the Works:

[*Insert the relevant portions and names of proposed sub-contractors*]

If you will give me your written confirmation that, notwithstanding the sub-letting of any portion of the Works, you will remain wholly responsible for the carrying out and completion of the Works in every respect in accordance with the contract, I will give my consent to such sub-letting.

Yours faithfully

Effectively, therefore, the sub-contractor requires the architect's consent, through the contractor, before unfixed materials can be moved. The consents must not be withheld unreasonably.

Clause 3.6.2.1.1 provides that the materials will become the property of the employer if the contractor has been paid under a certificate which includes the value of such materials. The sub-contractor is not to deny that the materials are the employer's property. Note that this provision takes no account of the fact that the contractor may not have paid the sub-contractor. If, unusually, the contractor does pay the sub-contractor before it has been paid itself, clause 3.6.2.1.2 provides that the materials become the property of the contractor. The purpose of clause 3.6.2.5 is presumably to safeguard the employer's rights under clause 4.12.2.1 to acquire ownership of the off-site materials through the contractor. That is because the contractor can pass rights to the employer only if those rights are already owned by the contractor. When the contract says that property in an item 'vests' in the contractor, it means, in simple terms, that the contractor owns that item. In this instance, the rights refer to the ownership of the materials.

A similar set of provisions is included in SBC. They were originally prompted by a case in which the employer paid for sub-contractor materials, but, before the contractor paid the sub-contractor, the contractor went into liquidation. The goods were held to be the property of the sub-contractor, and the employer had to pay for them again. Whether the current provisions are effective in preventing a recurrence of this problem, it is difficult to say. It may be expected that sub-contractors will increase their prices to cover the risk that they now appear to take.

However, it must be emphasised that these provisions will not necessarily protect the employer. For example, if the sub-contractor's materials have been supplied on terms of sale including a retention-of-title clause, this would not be defeated by clause 3.6.2.1.1 or any corresponding provision in the sub-contract. Clause 3.6.2.2 provides for the stepping down into the sub-contract of the rights of access for the architect to sub-contract work.

Clause 3.6.2.3 provides that the contractor must include in the sub-contract a specific term to make the contractor liable to pay simple interest to the sub-contractor if the contractor fails to pay the amount due to the sub-contractor by the final date for payment. The rate of interest is to be 5% (stated in the definitions, clause 1.1) over the base rate of the Bank of England, echoing the provisions for the contractor in IC clauses 4.8.5 and 4.14.5. Payment of interest is not to be taken as a waiver of the sub-contractor's right to payment in full at the correct time. This provision is to maintain the sub-contractor's entitlement to take action as set out elsewhere in the sub-contract. There is no need, as formerly, to make express reference to the sub-contractor's right to suspend work and to terminate its own employment. The payment of the interest is to be treated as a debt due from the contractor to the sub-contractor. A new provision (clause 3.6.4) covers the provision by the sub-contractor of a warranty which corresponds to the relevant warranty in the contract documents. The sub-contractor is to have 14 days to do this. It must be remembered that the provisions of a contract cannot bind anyone who is not a party to it. Therefore this clause places the onus on the contractor to ensure that an appropriate term is included in the sub-contract.

8.2.2 Named persons as sub-contractors

These provisions (clause 3.7 and schedule 2) are to be found only in IC. They are also quite complicated and, in places, unclear. A named person may be involved in one of two ways. One is if work is included in the contract documents to be priced by the contractor and carried out by a named person. In this case, the person is named in the contract documents, and the contractor does not have a choice. The second is if work is included in an architect's instruction regarding the expenditure of a provisional sum, and a person is named to carry it out.

The consequences are slightly different. The flowchart in Figure 8.3 outlines the procedures. In the first instance the contractor has 21 days from entering into the main contract to enter into a sub-contract with the named person. That is 21 days from the date the contract was formed, which may be 21 days from the date of the employer's unequivocal acceptance of the tender, not 21 days from signing the contract documents. It is a very short period. The sub-contract must consist of the Intermediate Named Sub-Contract Agreement ICSub/NAM/A (this incorporates sub-contract con-ditions ICSub/NAM/C).

These documents are clearly intended by the third recital to be part of the documents given to the contractor when inviting the tender and to be made part of the contract documents (see definition in clause 1.1). It is obviously envisaged that all such sub-contract tenders will be invited before the main contract is let. Introducing named sub-contractors is certainly not a soft option for architects who are short of the time necessary to produce full drawings and specifications.

Figure 8.3
Named sub-contractors (clause 3.7 and schedule 2)

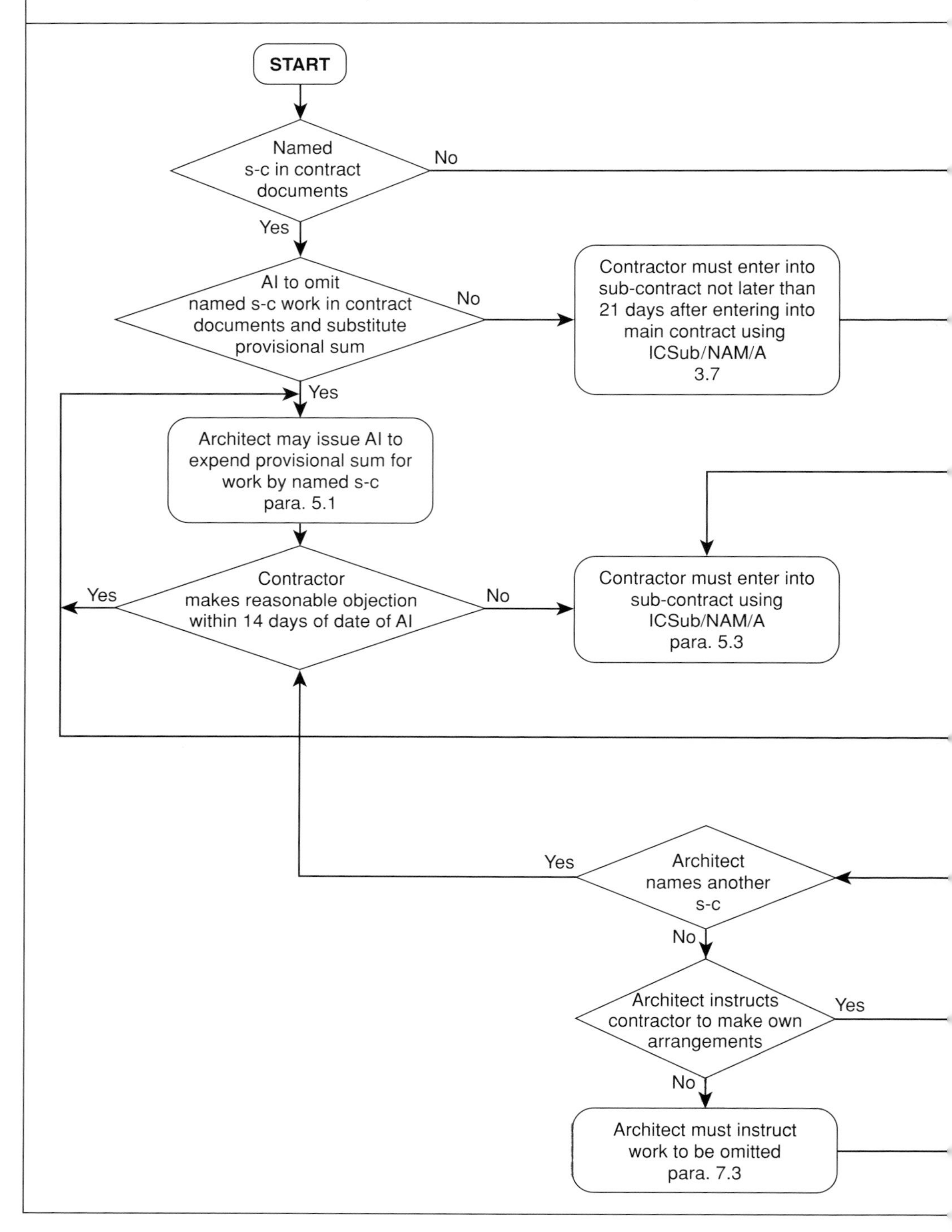

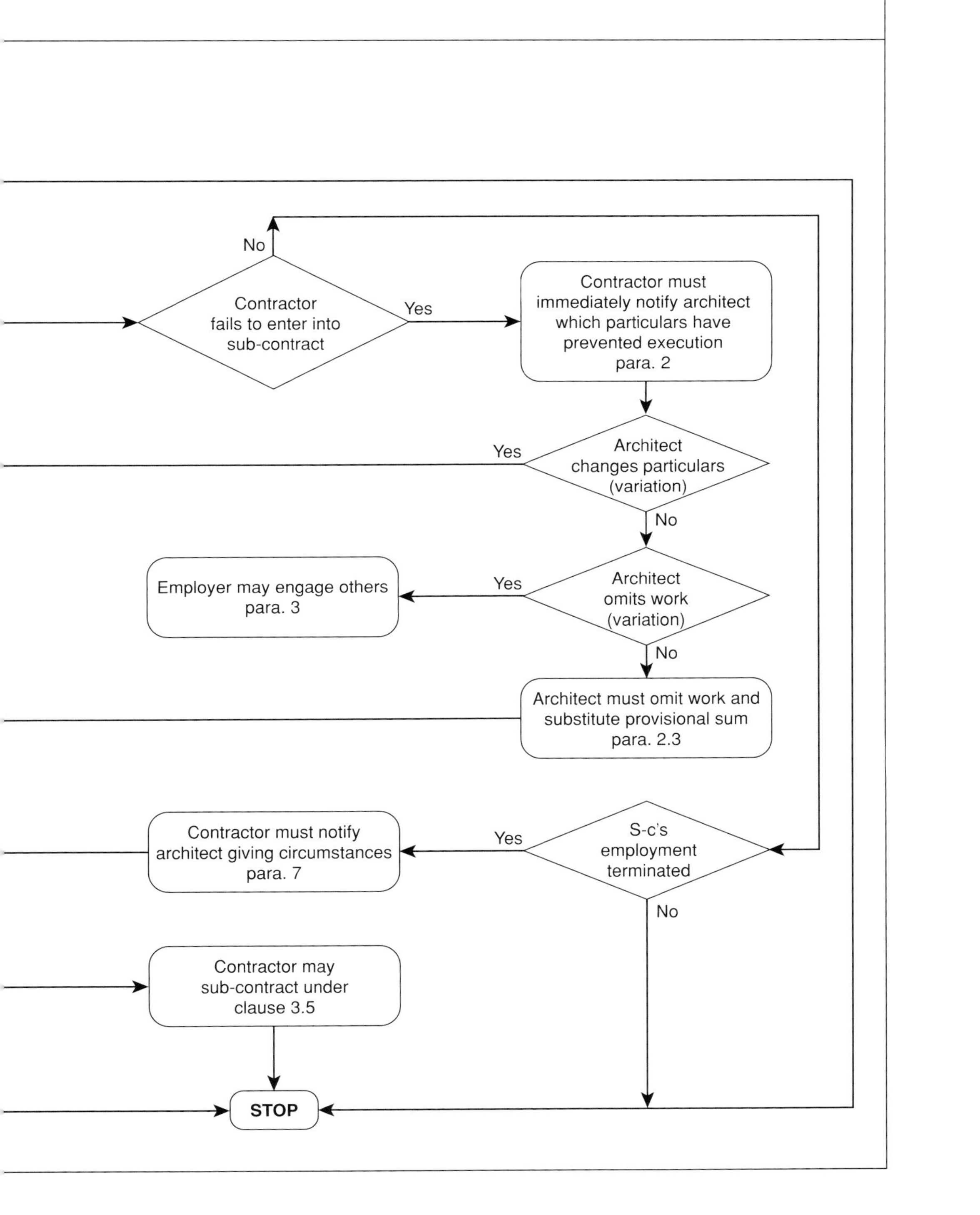
No
Contractor fails to enter into sub-contract
Yes
Contractor must immediately notify architect which particulars have prevented execution para. 2
Yes
Architect changes particulars (variation)
No
Employer may engage others para. 3
Yes
Architect omits work (variation)
No
Architect must omit work and substitute provisional sum para. 2.3
Contractor must notify architect giving circumstances para. 7
Yes
S-c's employment terminated
No
Contractor may sub-contract under clause 3.5
STOP

Paragraph 2 of schedule 2 provides that, if the contractor is 'unable to enter into a sub-contract in accordance with clause 3.7 and the particulars given in the Contract Documents', the architect must immediately be informed, specifying which particulars have caused the problem. 'Particulars' is the invitation to tender and tender (ICSub/NAM/IT and ICSub/NAM/T) together with the tender documents referred to therein. If the architect is reasonably satisfied that the particulars specified have indeed prevented the execution of the sub-contract (presumably this requires the contractor to send a letter from the sub-contractor stating as much), the architect has three courses:

- Alter the particulars to remove the sub-contractor's objection, which creates a variation (paragraph 2.1); or
- Omit the work altogether, which also creates a variation (paragraph 2.2); or
- Omit the work from the contract documents, substituting a provisional sum for which an instruction must be issued under paragraph 5 (paragraph 2.3).

This clause might appear to give the architect the power to vary the terms of the contract, which clearly only the employer can do with the consent of the contractor. It must therefore be assumed that such power is limited under paragraph 2.1 to the varying of other than contract terms. If the work is omitted altogether, the employer may get someone else to do the work and pay direct, subject to the provisions of clause 2.7 (see section 8.4).

The contractor must notify the architect of the date when a sub-contract has been executed, assuming that all goes well (see Figure 8.4). Before the contractor does so, the architect may issue an instruction similar to paragraph 2.3 (paragraph 4).

Paragraph 5 deals with the procedure if a named person arises through a provisional sum. This can occur in three ways:

- If the architect issues an instruction (paragraph 2.3) after the particulars are said to be preventing execution of the sub-contract; or
- If the architect issues a similar instruction (paragraph 4) omitting the work from the contract documents and substituting a provision sum before the sub-contract is signed; or
- If there is a provisional sum in the contract documents (clause 3.13). This clause allows the architect to issue an instruction regarding the expenditure of a provisional sum and to require that the work is to be done by a named person. The named person is to be employed by the contractor as a sub-contractor (paragraph 5.1).

Figure 8.4
Contractor to architect if contractor enters into sub-contract with named person

Dear

In accordance with paragraph 1 of schedule 2 of the contract, we hereby inform you that we entered into a sub-contract with [*insert name*] on the [*insert date*].

Yours faithfully

The instruction must describe the work to be done and include all the details of the invitation to tender and the tender (ICSub/NAM/IT and ICSub/NAM/T) with the documents referred to therein (paragraph 5.2).

Clearly there is a difference between a person being named in the contract documents, when the contractor has the opportunity to see who it is before tendering for the whole contract, and being named in an architect's instruction, when the job is in progress and the contractor is committed. For that reason, the contractor has 14 days from the date of the issue of the instruction in which to make an objection (see Figure 8.5). The objection must be reasonable. The contract does not say what is to happen if the contractor does make a reasonable objection. Presumably, the architect must name another person and so on, if the contractor continues to object, until the contractor stops objecting.

Under such circumstances the naming of a person is not a quick process. Whether or not the contractor's objection is reasonable may be referred to adjudication under clause 9.2. The architect has to complete the invitation to tender part of ICSub/NAM of the form of tender and agreement and send a copy to each of the persons it is desired to ask to tender. Each tenderer must complete the tender part and return it. Assuming that the architect has an inexhaustible supply of suitable tenderers, the problem is the delay to the contract. It is unclear whether the architect has power to award an extension of time. Clause 2.20.2.2 refers to compliance with architect's instructions 'to the extent provided therein', under clause 3.7 and schedule 2.

There is no provision for an extension of time for delay caused by the contractor's making reasonable objection. It is likely that the architect will treat the delay as stemming from the original instruction, and that seems to be a sensible approach. Alternatively, the work may be postponed under clause 3.12 and an extension given under clause 2.20.2.1. The contractor will then claim loss and/or expense under clause 4.18.2.1.

When the contractor's objections are sorted out, or if there are no objections, it must enter into a sub-contract with the named person, using the agreement part of ICSub/NAM. If the contractor is unable to enter into a sub-contract because of the 'particulars', the contract is again silent regarding the next move. What is certain is that the employer has the responsibility of naming a person who will enter into a sub-contract on the basis of the particulars. The best way to settle this situation is by negotiation. The architect should attempt to come to some agreement about the particulars with all parties. If that fails, another person must be named. The provisions apparently

Figure 8.5
Contractor to architect making objection to named person

Dear

We are in receipt of your instruction number [*insert number*], dated [*insert date*] instructing us to enter into a sub-contract with [*insert name*].

We have reasonable objection, under paragraph 5.3 of schedule 2 of the contract, to entering into such sub-contract. The reason for our objection is [*explain*].

Yours faithfully

work reasonably well in practice despite any lack of clarity in the drafting.

The employer can always arrange to amend this clause, as any other, at tender stage, but it should be done only with expert advice. It is essential that the employer enters into the ICSub/NAM/E agreement between the employer and a person to be named as a sub-contractor under the JCT Intermediate Building Contract. This will enable the employer to take direct action against the named sub-contractor in case of default, in particular if there is a failure in any of the following:

- Any design of the sub-contract works that the sub-contractor has undertaken to do
- The selection of the kinds of materials for the sub-contract works that the sub-contractor has undertaken to do
- The satisfaction of any performance specification for the sub-contract works that the sub-contractor has undertaken to fulfil

– all of which collectively might be referred to as 'design'.

Paragraph 11 of IC expressly removes the contractor's liability for the above items 'whether or not' the named sub-contractor is responsible to the employer for them. Moreover, the named sub-contractor is not to be liable for them through the contractor. The only way the employer can obtain any contractual redress is if an ICSub/NAM/E form has been executed. It is conceivable that action could be taken against the sub-contractor for negligence, but in the current legal climate it is unlikely to meet with success: *Murphy* v. *Brentwood District Council* (1990). Even with ICSub/NAM/E in place, the employer will be unable to recover directly from the sub-contractor if loss is caused by the carrying out of the sub-contract works (*Greater Nottingham Co-operative Society* v. *Cementation Piling and Foundation Ltd* (1988)), and a suitable amendment to the warranty is required.

No other sub-contractor is to be responsible for design through the contractor, and in that case there is no way in which the employer can obtain redress, because there is no privity of contract between the employer and the contractor's domestic sub-contractor under clause 3.5. In the case of some domestic sub-contractors there is an element of design in their work, and it may be sensible for the employer to enter into a form of warranty to establish a contractual route for redress in the case of design failure.

The phrase in this clause is that neither the contractor nor the named sub-contractor is responsible under the contract 'for anything to which such terms relate'. The precise meaning of the phrase may be a fruitful source of dispute. The clause, however, does not affect the contractor's normal obligations or those of any sub-contractor in the supply of goods and materials and workmanship. The previous edition of this form expressly excluded any 'other sub-contractor' from responsibility 'through the Contractor' for design. It appears that JCT has now quite sensibly taken the view that, if the contractor is not responsible for design, no sub-contractor can be held responsible 'through' the contractor. Any redress against a sub-contractor would have to be by means of a separate warranty agreement.

There are extensive provisions to deal with the termination of the named sub-contractor's employment under the sub-contract and its consequences. The contractor must advise the architect of any events which are likely to lead to termination under the sub-contract. The contractor must do this 'as soon as is reasonably practicable': in other words, the architect must be informed as soon as the contractor finds out (Figure 8.6). What the architect does then is not stated. Presumably, efforts will be made to find a solution short of termination. As the sub-contractor has no contractual relationship likely to give the architect any real influence, it is difficult to see what the architect might do.

If termination takes place, whether or not the architect has been advised of the likelihood, the contractor must write, giving the circumstances. The architect must then issue such an instruction as may be necessary, which must be one of the following:

- Name another person to do the work or the balance of the work, incorporating a description of the work and particulars of the named persons, Tender in ICSub/NAM/IT and ICSub/NAM/T with all the documents referred to therein, subject to the contractor's reasonable objection within 14 days as in paragraph 5.3 (paragraph 7.1)
- Instruct the contractor to make its own arrangements to do the work, either itself or by sub-contract under clause 3.5 (paragraph 7.2)
- Omit the work still to be finished (paragraph 7.3).

If the architect omits the work, the employer may, under clause 2.7, arrange to have the work done by others and pay direct.

The contract highlights a difference in consequence, depending on whether the work was originally included in the contract documents or is the result of an instruction regarding the expenditure of a provisional sum:

Figure 8.6
Contractor to architect if termination of the named sub-contractor's employment possible

Dear

In accordance with paragraph 6 of schedule 2 of the contract, we have to advise you that the following events are likely to lead to the termination of [*insert name*]'s employment:

[*Describe events*]

We should be pleased to receive your instructions as a matter of urgency.

Yours faithfully

- If work originally included in the contract documents: a paragraph 7.1 instruction ranks as a clause 2.20 relevant event (extension of time) but not as a clause 4.18 relevant matter (disturbance of progress). The contract sum must be increased or reduced to take account of the price of the second as compared with the first named sub-contractor. Amounts included in the second price for repair of the first named sub-contractor's defective work must be excluded from the contract sum. Thus the contractor is not to be held financially responsible for the termination, but it is held responsible for any defects at that time existing. A paragraph 7.2 or 7.3 instruction ranks as a clause 2.20 relevant event (extension of time) and as a clause 4.18 relevant matter (disturbance of progress). It also ranks as a variation
- If the result of an instruction regarding the expenditure of a provisional sum: a paragraph 7.1, 7.2 or 7.3 instruction ranks as a clause 2.20 relevant event (extension of time) and as a clause 4.18 relevant matter (disturbance of progress). It also ranks for payment as a further instruction under the provisional sum.

There is an important proviso that, if the instruction was issued as a result of some default of the contractor, none of the above benefits apply to the contractor. This is so whether the work was originally included in the contract documents or is the result of an instruction regarding the expenditure of a provisional sum. It is only fair to stress that the above is intended to clarify what is regarded as an unhappily worded paragraph. A strict reading of paragraph 9, for example, may make the employer liable to pay the contractor for the sub-contract work if the contractor is in default and a 7.3 instruction omitting the work is issued. That cannot be what was intended.

The contractor must take whatever action is necessary, within reason, to recover from the named sub-contractor any additional amount that the employer has had to pay to the contractor due to the issue of the instructions under paragraphs 7.1, 7.2 or 7.3 and the consequences thereof, and any liquidated damages that the employer would have been able to recover from the contractor had it not been for the instructions. There is no time limit set for the contractor to take action. Presumably, it will take the action after practical completion; otherwise, how is the contractor to know the amount of liquidated damages which would have been recoverable by the employer (paragraph 10.2)?

The contractor is required to take action only where the employment of the sub-contractor has been terminated under ICSub/

NAM/C clauses 7.4, 7.5 or 7.6. There is a proviso that the contractor cannot be required to commence any arbitration or other proceedings unless the employer agrees to pay any legal costs the contractor incurs. It is difficult to see what action the contractor can take (unless some of the named sub-contractor's money is being held) if the employer decides against paying legal costs. The contractor is to account – i.e. presumably prepares an account – for any amounts recovered. Insofar as the contractor fails to take action, it is liable to the employer for the additional amount, including the amount equal to liquidated damages, payable by, or due to, the employer.

Whether or not the employer agrees to fund legal proceedings, it is difficult to understand, assuming that insufficient funds are retained to allow set-off, how the architect can demonstrate that the contractor has failed to take action provided the contractor has written some sternly worded letters to the named person (see Figure 8.7). The provision for named persons has been slightly reorganised since the last edition, but the wording is much the same as before and still quite complex.

8.3 *Statutory authorities*

Certain crucial parts of most, if not all, contracts are carried out by local authorities or statutory undertakers such as the gas and electricity suppliers. Where they carry out the work solely as a result of their statutory rights or obligations, they are not to be considered named sub-contractors (schedule 2, paragraph 12). The implication is that, where they carry out work which is not a result of their statutory rights or duties, but as a matter of contract, they may be named sub-contractors. They may also, in such instances, be considered as sub-contractors to the contractor in the traditional way (clause 3.5) or persons employed by the employer to carry out work outside the contract (clause 2.7; see section 8.4) depending on the circumstances of their engagement. In carrying out their statutory duties, the undertakers have no contractual liability, although in some cases they have tortious liability. When they are carrying out work outside their statutory duties, they are exactly like anyone else who enters into a contract. If a supplier of electricity, for example, delays the completion of the works by carrying out (or failing to carry out) work in pursuance of its statutory duty, the contractor will be entitled to an extension of time (clause 2.20.7), but not for delay caused by other work that the electricity supplier carries out or fails to carry out in its capacity as sub-contractor to the main contractor.

Figure 8.7
Contractor to named sub-contractor to recover amounts under paragraph 10.2.1 of schedule 2 of the main contract

Dear

In accordance with clause 7.7.4 of the sub-contract, we require you to pay us the sum of [*insert amount*], which we are obliged under paragraph 10.2.1 of schedule 2 of the main contract to recover from you.

If we do not receive the above mentioned sum by [*insert date*], we may take adjudication or other proceedings against you.

Yours faithfully

The contractor must comply with all statutory requirements: for example, the Planning Acts and dependent regulations and the Construction (Design and Management) Regulations 1994. The contractor is responsible for giving any notices required and for paying any fees or charges in connection with the Works (clause 2.3), for which it is entitled to be reimbursed by having the amounts added to the contract sum, unless they are already included in the contract documents. Thus, if the appropriate contract document is a specification and it states that the contractor must allow for paying all statutory fees and charges, the contractor must include the amount in the tender price.

The contractor is not liable to the employer if the Works do not comply with statutory requirements if this is because they have been carried out in accordance with the contract documents or any of the architect's instructions (clause 2.15.3). This provision will not, of course, affect any liability the contractor may have to the appropriate statutory undertaking or to the local authority. There is a proviso that, if the contractor finds any divergence between the statutory requirements and the contract documents or architect's instructions, the divergence must be specified to the architect in writing (clause 2.15.1). It may be small comfort to the architect because, if the contractor does not find a divergence which exists, it is not liable. It is the architect's duty to provide correct information: *London Borough of Merton* v. *Stanley Hugh Leach Ltd* (1985). If an emergency arises, the contractor must comply with statutory requirements without waiting for the architect's instructions. It must carry out and supply just enough work and materials as are necessary to comply as an immediate measure, and inform the architect forthwith, not necessarily in writing. A contractor which satisfies these requirements is entitled to be paid as though it has carried out the architect's instructions requiring a variation, provided that the emergency arose because of a divergence between the statutory requirements and the contract documents and/or any of the architect's instructions, drawings or documents referred to in clauses 2.13.1 or 2.11(clause 2.16.3).

Thus, if the emergency arises through the contractor's default or inefficiency, it is not entitled to payment except insofar as the architect may be prepared to accept that there would have been some additional cost irrespective of the contractor's default.

Clauses 3.18 and 3.19 require the parties to comply with the CDM Regulations. By making compliance with the Regulations a contractual duty, breach of the Regulations becomes a breach of contract, so providing both employer and contractor with remedies under the contract. Clause 3.18.1 provides that the employer 'shall ensure' that the planning supervisor carries out all the relevant duties under the Regulations and that, where the principal contractor is not the contractor, it will also carry out its duties in accordance with the Regulations. There are also provisions that the contractor, if it is the principal contractor, will comply with the Regulations and 'ensure' that any sub-contractor will provide the information to the planning supervisor or to the principal contractor as appropriate which the planning supervisor will reasonably require for the preparation of the health and safety file (clauses 3.18.2 and 3.18.3).

Every instruction which the architect issues potentially carries a health and safety implication, which must be examined and the appropriate procedural steps taken under the Regulations. The Regulations present the planning supervisor with grave responsibilities. Some of his or her duties must be carried out before work is started on site. If necessary actions delay the issue of an architect's instruction or, once issued, delay its execution, the contract provides that the contractor is entitled to extension of time (clause 2.4.17) and any loss and/or expense it can substantiate (clause 4.12.9). Every architect should have a thorough grasp of the Regulations and the contractual clauses that deal with them so that the full consequences of any new instruction can be carefully considered before it is issued.

8.4 *Work not forming part of the contract*

The employer has the right to enter into contracts with persons other than the contractor to carry out work on the site (clause 2.7). Such persons are commonly firms or individuals over whom the employer wants complete control. For example, when the employer is a local authority, they may be the authority's own employees. Or the employer may have a special relationship with such persons as artists, sculptors, graphic designers, landscapers etc. and that is the reason for wanting to employ them in a direct way. In principle, the contractor should be responsible for all the work to be done. It promotes efficiency on site and removes areas of possible dispute. If the work is to be considered not to form part of the contract, it must be the subject of a separate contract between the employer and the person to provide the work, and it must be paid for by the employer directly to the person employed and not through the contractor.

Work carried out by statutory undertakers not in pursuance of their statutory duties may fall into this category.

The contract provides for two situations. The first is where the contract documents provide for such work by informing the contractor what is to be carried out, when, and (possibly but not necessarily) by whom. The documents should require the contractor to allow the work to be carried out as stated, and give details of any items of attendance which may be required. The contractor should then make provision for the work in the programme. The second situation is where the contract documents do not provide for such work.

In the first situation, the contractor must allow the work to be carried out and must provide whatever attendance is specified and priced for. In the second, the employer must first obtain the contractor's consent to the proposals before arranging to have the work carried out. The contractor must not withhold its consent unreasonably. It would be reasonable to withhold consent if the proposed work would constitute a severe disturbance to the contractor's progress (Figure 8.8). It would be unreasonable if the work was to be done without affecting the contractor's activities in any way. Between these two extremes lie many situations which may require the exercise of goodwill on both sides.

Whether the work not forming part of the contract is provided for in the contract documents or is simply the subject of the contractor's consent does not affect the employer's responsibility under the contract. For the purposes of the insurance section 6, persons directly employed by the employer are deemed to be employer's persons. They are not to be deemed sub-contractors. The result is that the employer may have uninsured liabilities. The employer should therefore obtain the necessary cover through an insurance broker. The broker should be given a copy of the insurance clauses and be requested to arrange cover for the employer, and those for whom the employer is responsible, in respect of any act or neglect of those persons. This is the employer's responsibility, not the responsibility of the architect. The directly employed persons may already have adequate insurance cover, but it is a matter best left in the hands of a broker with experience in this kind of insurance.

The contractor is entitled to be awarded extensions of time if completion is delayed by the carrying out (or failure to carry out) of work not forming part of the contract. This now falls under the general prevention or default clause 2.20.6. It is a very easy claim to make, and a contractor running late on the contract will usually be delighted to give its consent, if required, to any work by the employer's directly employed persons. They can then be made to appear responsible for any delays thereafter. Therefore the architect must be especially vigilant when examining such claims; a network analysis is invaluable.

Figure 8.8
Contractor to employer, objecting to work to be done by employer's directly employed contractors

Dear

We understand that you wish to arrange for [*insert brief description of the work*] to be carried out by other persons under the terms of clause 2.7.2.

We consider that it is reasonable for us to withhold our consent because [*give brief reasons*].

Yours faithfully

Copy: Architect

The employer's responsibility to supply, or a failure to supply, those materials which the employer has agreed to supply is closely linked to the problem of work outside the contract. In both cases the contractor is not involved with the external contracts, and it may be thought that the employer is assuming needless responsibility. Suppose that the employer wishes to supply all the paint for a contract, perhaps because a special rate is available. In order to avoid any claims from the contractor, the employer must supply paint of the correct colours, in the correct quantities, of the correct types (undercoat etc.), and at the time it is required. Moreover, any unsatisfactory paint may provoke a claim.

The wording of the clause relating to extensions of time, discussed above, is precisely repeated as relevant matters for which the contractor may make a claim for loss and/or expense (clause 4.18.5). The employer is extremely vulnerable, and the architect must exercise great care in deciding whether the claim is valid.

Potentially the most damaging result of the employer's directly employing persons to carry out work outside the contract, or of arranging to supply materials, is that the contractor may acquire grounds to terminate its employment (clause 8.9.2.2). The matter is dealt with in section 12.2.2. Although the contractor's potential remedies of extension of time, loss and/or expense and termination of employment may appear severe, they are in truth merely a reflection of the problems caused to the contractor by introducing other contractors or suppliers on the site over whom the contractor has no control.

If the employer expresses an intention of directly employing operatives or of supplying any materials, the architect has a duty to give advice about the pitfalls. Most of them can be avoided by ensuring that all work is done, and all materials are supplied, through the contractor. In view of the possibility of named sub-contractors in this contract, the employer would be unwise to arrange for work to be carried out which does not form part of the contract.

8.5 *Third party rights and collateral warranties*

Although under SBC, the JCT has taken a brave step in making use of the Contracts (Rights of Third Parties) Act 1999 in order to exclude the need for certain independent warranties, IC and ICD continue to exclude its effect entirely in clause 1.6.

However, clauses 7.2, 7.3, 7.4, 7.5, 7.6, 7.7 and 7.8 deal with contractor's warranties for purchasers and tenants and funders, and sub-contractors' warranties for purchasers and tenants/funders and the employer. The system depends on the relevant details being inserted in the contract particulars. There is a sensible proviso in clause 7.4 that, whether the contract is executed as a deed or under hand, the warranty will be executed in the same way. This avoids the difficulties sometimes encountered in the case of bespoke warranties when the limitation periods do not accord with the main contract.

Clause 7.3 stipulates that any notices to be given under clauses 7.5 to 7.8 must be in writing and sent by actual, recorded or special delivery. The criteria for receipt of such notices are to be the same as set out in clause 8.2.3 in the termination provisions. Where the warranty to be used is other than the specified JCT edition, a copy of the warranty must be sent with the notice. It is a basic contract principle that a party is bound only by the obligations that are included in the contract. They can be incorporated by reference provided the reference is clear and unambiguous. Reference to the contractor completing a JCT warranty identified by its normal reference is obviously sufficient, because such a warranty is well known to the industry as a whole. But if a different warranty is to be used, it is essential that a copy of that warranty be included in the contract documents; otherwise the contractor cannot be obliged to execute it. Therefore it is not sufficient that a copy of the warranty is sent with the notice; that alone will not bind the contractor.

Clauses 7.5 and 7.6 deal with warranties to purchasers and tenants, and funders respectively. Where purchasers or tenants are identified in part 2 of the contract particulars, the employer may send a notice to the contractor requiring the contractor, within 14 days of receipt, to execute a warranty with the relevant persons in form CWa/P&T. The warranty is to be completed in accordance with part 2 (B) of the contract particulars.

Part 2 (A) contains space for the identities of the purchasers and tenants to be inserted, together with a description of the part of the Works to be purchased or let. Note that it is not necessary for the name of the purchaser or tenant to be inserted. It is highly likely that the specific names will not be known at the date the contract documents are being prepared. What must be inserted is the name or the class of person or a description. The important point is that it should be unmistakable which person is being identified. Failure to insert this information would prevent a warranty from being completed.

(B) sets out how various clauses in the warranty are to be completed and includes reference to maximum liability and net contribution.

The arrangements for warranties for funders are virtually identical. The warranty is CWa/F, and it is to be completed in accordance with (D) of part 2 of the contract particulars. Of importance in the case of both warranties is the recording of the identities of additional consultants and any relevant sub-contractors.

Where sub-contract warranties are concerned, the position is more complicated. Clauses 7.7 and 7.8 refer. The sub-contractors who are to give warranties are to be listed in part 2 (E). Within 21 days of receipt of a notice from the employer, listing the sub-contractor, type of warranty and the beneficiary, the contractor must obtain the warranties in the forms SCWa/P&T or SCWa/F as appropriate. If a bespoke warranty is to be used, a copy must be included in the contract documents, and there is provision for reference to it in the contract particulars, otherwise the contractor will not be bound by it.

That leads to another point. Clause 7.7 states that the obtaining of the warranties is subject to any amendments proposed by the sub-contractor concerned and approved by the employer and the contractor. There is the usual proviso that the approvals are not to be unreasonably delayed or withheld. It appears, therefore, that although the sub-contractor can propose amendments, they are effective only if approved by the employer and the contractor. But, of course, the sub-contractor is not bound by anything in the main contract IC. Therefore whatever is included in IC about sub-contractor warranties will not have any binding effect on the relevant sub-contractors. In order to ensure that the sub-contractor is bound, the requisite information must be included in the relevant sub-contracts.

Clause 7.8 deals with warranties to be given by sub-contractors to the employer. The situation is similar to other warranties given by sub-contractors under clause 7.7. No form of warranty is specified, and it is for the employer to attach a copy of the preferred warranty as noted above. This list of sub-contractors must not include any named sub-contractors, because there is a special form of warranty for use with a named sub-contractor.

Especial care must be taken in completing part 2 of the contract particulars.

8.6 *Summary*

Assignment and sub-contracting

- Neither party may assign without the other's consent
- The contractor may sub-let with the architect's consent

- The sub-contractor's employment terminates on the termination of the contractor's employment
- The contractor must consent to the sub-contractor removing unfixed materials from site
- Unfixed materials become the employer's property when the contractor is paid
- Unfixed materials become the contractor's property if it pays the sub-contractor before being paid itself.

Named sub-contractors

- Named sub-contractors may arise by inclusion in the contract documents or by an instruction regarding a provisional sum
- ICSub/NAM must be completed by the architect, the named sub-contractor, and the contractor
- The work may be omitted and carried out by the employer's own men
- The named sub-contractor in the contract documents may be replaced by a named sub-contractor in an architect's instruction
- The contractor may object to any named sub-contractor in an instruction
- The employer should enter into the ICSub/NAM/E form of agreement with the named sub-contractor
- Design, selection and satisfaction are not the liability of the contractor, nor are they the liability of the sub-contractor through the contractor
- On termination of a named sub-contractor's employment, the architect may name another person, request the contractor to make its own arrangements, or omit the work
- Under certain circumstances, extension of time and loss and/or expense may be awarded to the contractor after termination
- The contractor must take action to recover the employer's losses after termination of the named sub-contractor's employment
- If the contractor fails to take appropriate action, it is liable for the losses.

Statutory authorities

- In pursuance of their statutory duties, statutory undertakers are not liable in contract but may give grounds for extension of time
- Not in pursuance of their statutory duties, they are liable in contract and may be sub-contractors, named sub-contractors or employer's persons

- The contractor must comply with statutory requirements
- The contractor is not liable to the employer if it works to contract documents or instructions, provided that it notifies any divergence found
- The contractor may carry out and be paid for emergency work without instruction
- The contractor must notify the architect forthwith if there is an emergency
- The contract requires compliance with the CDM Regulations
- Compliance may give rise to extensions of time and claims for loss and / or expense.

Work not forming part of the contract

- The employer may directly employ people on site
- The contractor's consent is required if the work is not mentioned in the contract documents
- There are insurance implications
- The employer is vulnerable to claims for extension of time and loss and/or expense
- There are grounds for termination if work is delayed for a month by work not forming part of the contract.

CHAPTER NINE
POSSESSION, PRACTICAL COMPLETION AND DEFECTS LIABILITY

9.1 Possession

9.1.1 General

Possession is the next best thing to ownership. If the owner of a motor-car lends it to a friend, the friend has a better claim to the car than anyone else except the owner. The builder in possession of a site can, in general terms, exclude everyone from the site except (and often including) the owner. In practice, there are exceptions to this general rule, laid down by the building contract and by various statutory regulations.

A contractor carrying out building Works is said to have a licence from the owner to occupy the site for the length of time necessary to complete the Works. The owner has no general power to revoke such a licence during the contract, but it may be brought to an end if the contractor's employment or the contract itself is lawfully brought to an end.

If there were no express term in the contract giving the contractor possession of the site, a term would be implied that the contractor must have possession in sufficient time to allow it to complete by the contract completion date.

There is now an express term in IC allowing access to the Works at all reasonable times for the architect and any person authorised by the architect. This will usually mean other people from the architect's practice, consultants and the clerk of works. It is suggested that authorisation implies that the architect will give written notice to the contractor listing the relevant persons.

9.1.2 Date for possession

The date, or dates if sections are used, for possession is to be entered in the contract particulars. Clause 2.4.7 states that, on that date, possession must be given to the contractor, and that, subject to any extension of time which may be awarded, the contractor must proceed regularly and diligently with the Works and complete them on or before the date for completion in the contract particulars. The phrase 'regularly and diligently' is discussed in section 5.1.2.

Under the general law, if the employer fails to give possession on the due date, it is normally a serious breach of contract. By 'possession of the site' is meant possession of the whole site and not just a part unless sections are used: *Whittal Builders* v. *Chester-Le-Street District Council* (1987) (the second case). The contractor would have a claim for damages at common law, and the time for completion would become 'at large'. That is to say, the contractor's obligation would be to complete the Works within a reasonable time, and no date could be established from which liquidated and ascertained damages begin to run: *Peak Construction (Liverpool) Ltd* v. *McKinney Foundations Ltd* (1970); *Rapid Building Group Ltd* v. *Ealing Family Housing Association Ltd* (1984).

The problem is partially overcome in IC by the inclusion of an optional clause (2.5), which permits the employer to defer giving possession for a time which must not exceed a period to be inserted in the contract particulars. From the employer's point of view the clause should always be included for protection. The contract recommends that the period of deferment should not exceed six weeks. The period is somewhat arbitrary because, at the time of tender – of signing the contract even – presumably the employer fully intends to give possession on the date stated in the contract particulars. If it was known at that time that the date was going to be deferred, the date in the contract could be adjusted accordingly.

It is perfectly possible to amend the contract so that the permitted deferment is, say, 12 weeks, but to introduce so large an element of uncertainty into the contract would almost certainly result in increased tender prices. If possession is delayed it is usually delayed for a relatively short time (because demolition contractors have not finished their work, or because planning or building regulation permission is delayed, or for some other similar cause). Otherwise, the delay is caused by some major problem which lasts for a considerable period, and in such circumstances it would be unfair to rely on a clause permitting deferment. The best-laid plans can go wrong, so although the architect may, in consultation with the employer, decide to insert a period shorter than six weeks, an insertion should not be omitted altogether.

There is no prescribed form of notice for deferment, but it must come from the employer. The architect would normally draft a suitable letter, which need not give any reason (Figure 9.1).

Figure 9.1
Employer to contractor, deferring possession of the site

Dear

[*If length of deferment is known*]

In accordance with clause 2.5 of the conditions of contract, take this as notice that I defer giving possession of the site for [*specify period*]. You may take possession of the site on [*insert date*].

[*If length of deferment is not known*]

In accordance with clause 2.5 of the conditions of contract, take this as notice that I defer giving possession of the site for a period not exceeding [*insert the period named in the contract particulars*]. I will write to you again as soon as I have a definite date for you to take possession.

Yours faithfully

Copy: Architect
Quantity surveyor
Consultants
Clerk of works

If possession is deferred, the contractor will be able to claim loss and/or expense and an extension of time.

Although failure to give possession on the due date is a serious breach of contract, IC now provides for the giving of an extension of time (clause 2.20.6) and loss and/or expense (clause 4.18.5) for a range of things under the general heading of 'impediment, prevention or default, whether by act or omission'. These catch-all clauses will clearly include a failure to give possession on the due date. They remove the need for a contractor to seek damages at common law for the breach if the failure to give possession extends beyond any deferment by the employer. Of course, despite the existence of such clauses, the contractor always has the option to pursue damages instead of or as well as using the contractual route, subject only to the proviso that there must be no double recovery.

The contractor normally gives up possession of the site at practical completion or on determination of its employment. It is, however, granted a restricted licence to enter the site for the purposes of remedying defects (see section 9.3).

The power to defer possession must not be confused with the power to order postponement of the work to be executed under the provisions of the contract. Loss and/or expense (clause 4.18.2.1) and extension of time (clause 2.20.2.1) can also be claimed for postponement. If the whole, or substantially the whole, of the works is suspended for the length of time stipulated in the contract particulars by postponement, the contractor may terminate its employment (clause 8.9.2.1). If there is postponement, the contractor has possession of the site, but the architect has suspended work. That is not to say that the contractor may not use some or all of the time the work is suspended to carry out work specifically connected with its occupation of the site (for example, repairing or improving site office accommodation, sorting materials, attending to security).

9.2 *Practical completion*

9.2.1 Definition

Clause 2.21 states that the architect must issue a certificate forthwith when, in his or her opinion, practical completion of the Works or a section is achieved and the contractor has complied sufficiently with clause 3.18.3. Clause 3.18.3 requires the contractor to provide and to ensure that any sub-contractor provides certain information. The sub-contractor, of course, is to provide it through the contractor.

The information in question, which must be information reasonably required to prepare the health and safety file required by the CDM Regulations, is to be provided to the planning supervisor. In cases where practical completion depends on this point alone, there are likely to be referrals to adjudication, to decide whether the planning supervisor's requirements are reasonable.

It is clear that the architect must issue one certificate to cover two events: practical completion in a physical sense, and the supply of health and safety information. The consequences of the certificate are considerable (see section 9.2.2). Despite that, the contract does not define the meaning of 'practical completion' in the physical sense. It is not the same as 'substantial completion', nor does it mean 'almost complete'. There is some useful case law, and the leading case is the House of Lords decision in *Westminster Corporation* v. *J. Jarvis & Sons Ltd* (1970). The point that emerges is that the architect is not to certify practical completion if any defects are apparent or if anything other than very trifling items remain outstanding: *H.W. Nevill (Sunblest) Ltd* v. *Wm Press & Son Ltd* (1981). Within these guidelines, the architect is free to exercise discretion. There will be differences of opinion on the question of what are 'trifling' items or what the courts refer to as *de minimis* items. Probably, what constitutes such items can fluctuate considerably depending on circumstances. A practical test to apply is whether the employer would be seriously inconvenienced while the *de minimis* items are being finished. If the employer would be inconvenienced, the architect is probably justified in withholding the certificate. When issuing a certificate, it is good practice to expressly exclude any such outstanding items: *Tozer Kemsley & Milbourne (Holdings) Ltd* v. *J. Jarvis & Sons Ltd and Others* (1983).

But there is no obligation on the architect to tell the contractor what items remain to be completed before the issue of the certificate. The temptation to issue lists of outstanding items should be resisted. The contractor knows what is required by the contract. The issue of lists at this stage is confusing and often leads to disputes. This is particularly the case, because such lists are often referred to as 'snagging lists' as though the items on such lists are something other than defects and breaches of contract. The onus of inspecting the work and preparing work lists for the contractor lies with the person-in-charge.

The contractor is not bound to notify the architect when practical completion has been achieved, but it is wise to do so, probably some weeks in advance (Figure 9.2). Some architects are in the habit of arranging so-called 'handover meetings' at which representatives of

Figure 9.2
Contractor to architect, notifying imminence of practical completion

Dear

We anticipate that the Works will be complete on [*insert date*]. Please confirm that you intend to carry out your inspection on that day. We are arranging for [*insert name*] to be on site to give immediate attention to any queries which may arise. We look forward to receiving your certificate of practical completion following your inspection.

Yours faithfully

Figure 9.3
Architect to employer if employer wishes to take occupation of the whole building before practical completion achieved

Dear

I refer to our meeting on site with the contractor on [*insert date*].

I confirm that, in my opinion, practical completion has not been achieved, so it is my duty under the provisions of the contract to withhold my certificate. However, I note that, for reasons of your own, you have agreed with the contractor to take occupation of the whole building. Although I think that you are unwise, as I explained on site, I respect your decision and I will continue to inspect until I feel able to issue my certificate. At that date, the rectification period will commence.

Yours faithfully

the employer and sometimes the employer's maintenance organisation are present. This can be a prudent move by the architect, on the principle that many eyes are better than one. There will also be consultants on hand to inspect their own particular portions of the Works. However, the decision to issue a certificate is solely the architect's. The responsibility cannot be moved onto the employer simply because the employer is present. The exception to that is if the employer insists that the building is ready, even though the architect's view that the building has not achieved practical completion is made clear to the employer. The employer may agree with the contractor to take possession of the building despite the architect's protests (see section 9.2.3). This often happens if an employer is anxious to get into a building, but the position is that the architect must not issue a certificate until the contractual requirements have been satisfied.

Occupation by the employer is not the same as practical completion, and, in the absence of a certificate of practical completion, it appears that the employer may still recover or deduct liquidated damages: *BFI Group of Companies Ltd* v. *DCB Integrated Systems Ltd* (1987); *Impresa Castelli SpA* v. *Cola Holdings Ltd* (2002). The architect will usually write to the employer and make the position clear (Figure 9.3). If the employer later discovers that outstanding items cause trouble, or if the contractor does not complete as quickly and efficiently as it promised, the employer will have been put on notice. The architect's duties will have been carried out properly with due regard for the employer's interests. The architect's duty to issue a certificate of practical completion will remain, but not until, in his or her opinion, the works have achieved that state.

9.2.2 Consequences

Clause 2.21 states that 'practical completion of the Works or the Section shall be deemed for all the purposes of this Contract to have taken place on the date stated in' the certificate. The 'purposes of this Contract' are to be found in various clauses throughout the contract:

- The contractor's liability for insurance under Schedule 1, option A ends
- Liability for liquidated damages under clause 2.23 ends (but see section 10.2.2)
- The employer's right to deduct full retention ends. Half the retention percentage becomes due for release within 14 days (clause 4.9)

- The six-month period begins during which the contractor must send all documents reasonably required for adjustment of the contract sum to the architect (clause 4.13)
- The period of final review of extensions of time begins (clause 2.19.3)
- The rectification period begins (clause 2.30).

9.2.3 Partial possession and sectional completion

There is often confusion between partial possession (clauses 2.25–2.29) and sectional completion. Sectional completion used to be dealt with in the form of a supplement involving a great many detailed changes to the contract. Sensibly, the new form uses wording which also applies to sections if the relevant parts of the contract particulars have been completed.

The essential difference is that partial possession may be exercised only if the contractor consents, even though that consent must not be unreasonably withheld or delayed. The employer having entered into the contract with one completion date, the decision to take partial possession is very much an afterthought. Sectional completion, on the other hand, establishes the employer's right to have the building completed in sections on specific dates, and the contractor's consent is not required. The sections will have been established at tender stage or, at the very latest, when the contract was executed.

Matters can become quite complicated if the project is divided into sections and the employer subsequently decides to take possession of part of one or more sections.

Note that the architect does not certify partial possession, but merely records it with a written statement which simply states which part or parts have been taken into possession by the employer, and the date when it occurred. This nicely separates the architect from any responsibility for the decision. In practice it is useful for the architect to incorporate a plan (and perhaps sections) suitably marked up to illustrate the part clearly.

Although partial possession is not the same as practical completion, clause 2.6 provides that practical completion is to be deemed to have occurred for two specific purposes:

- The start of the rectification period for that part and the subsequent issue of a certificate of making good (clause 2.27)
- The issue of a certificate under clause 4.9, releasing half the retention.

When parties *deem* something, they agree to act as though the thing deemed is true although they know it to be false: *Re Coslett (Contractors) Ltd, Clark, Administrator of Coslett (Contractors) Ltd in Administration* v. *Mid Glamorgan County Council* (1997). It has been held that, where partial possession has been taken of the whole of the Works under the partial possession clause, it amounts to practical completion: *Skanska Construction (Regions) Ltd* v. *Anglo-Amsterdam Corp Ltd* (2002). The judgment is hardly surprising. What is perhaps surprising is that an employer can purport to take partial possession of a whole.

If insurance options A or B apply, the obligation of the contractor or the employer comes to an end so far as the part taken into possession is concerned. Where the employer is insuring under option C, the obligation under paragraph C.1 to insure the existing structure must include the part taken into possession.

Clause 2.29 relates to liquidated damages. Its purpose is to provide for a proportionate reduction in the liquidated damages. The present clause is clear and unambiguous – an enormous improvement on the virtually incomprehensible clause which preceded it.

9.3 *Rectification period*

9.3.1 Definition

Clause 2.30 refers to defects liability. Reference is also made to the rectification period formerly known as the 'defects liability period'. The change in name appears to be change for change's sake. The length of the period is to be stated in the contract particulars. The period is for the benefit of all parties but principally for that of the contractor. The idea is to allow a specific period of time for defects to appear, list the defects, and give the contractor the opportunity to remedy them. Any defect is a breach of contract on the part of the contractor, which has agreed to carry out the work in accordance with the contract documents. If there was no rectification period, the employer's only remedy for defects would be to take action at common law. So the insertion of the period provides a valuable method of identifying defective work and having it corrected. Without it, the contractor would have no right or duty to return. This is a sensible provision aimed at keeping the costs of making good to a minimum.

Contrary to what many contractors (and some architects) appear to think, the contractor's liability for defects does not end at the end of the rectification period; what does end is its right to correct them (and even that right is limited, as will be seen). Afterwards, the employer is free to take legal action for damages if further defects appear, although in practice the employer will normally be satisfied if the defects are corrected.

Contractors commonly refer to the rectification period as the 'maintenance period'. This is misleading and wrong. Maintenance implies a far greater responsibility than simply making good defects, for example touching up scuffed paintwork and attention to general wear and tear. Even in other standard form contracts where the term is used (i.e. ACA 3 and GC/Works/1 (1998)), the actual wording of the clause usually restricts the obligation to making good defects.

9.3.2 Defects, shrinkages or other faults

The contractor is required to make good 'defects, shrinkages, or other faults'. At first sight this might appear to be all-embracing. In fact 'other faults' is to be interpreted *eiusdem generis*. That is to say that they must be faults which are similar to defects or shrinkages. A defect occurs when something is not in accordance with the contract. If an item is in accordance with the contract, it is not defective for the purposes of this clause. It might be less than adequate in some way, but that could be due to a fault in design, and therefore the architect's responsibility. Shrinkages are a source of dispute on many contracts. They become the contractor's liability only if they are due to material or workmanship not in accordance with the contract.

For example, shrinkage most commonly occurs in timber, caused by a reduction in moisture content after the building is heated. It is the architect's job to specify a suitable moisture content for the situation. If shrinkage occurs during the rectification period, it can only be because the timber was supplied with too high a moisture content, or because the architect's assumption about the appropriate moisture content was incorrect. Only the first explanation is the contractor's liability. In practice, it is often very difficult to decide which explanation applies, and the architect will naturally be drawn to the conclusion that the specification is correct. If the contractor objects (Figure 9.4), the only way to be sure is to have samples cut out and tested in the laboratory – a very expensive procedure. All too often the culprit is the occupier of the building, who is running the central heating above the recommended temperature. In such a case the architect's assumptions were, in effect, wrong, and the contractor is not responsible for making good.

Figure 9.4
Contractor to architect, objecting to some items on the schedule of defects

Dear

Thank you for your instruction number [*insert number*], dated [*insert date*], scheduling the defects you required to be made good now that the rectification period has ended.

We have carried out a preliminary inspection and we are making arrangements to make good most of the items on our schedule. However, we do not consider that the following items are our responsibility for the reasons stated:

[*List, giving reasons*]

Naturally, we shall be happy to carry out such items of work if you will let us have your written agreement to pay us daywork rates for so doing.

Yours faithfully

9.3.3 Frost

There is no longer an express reference to frost damage in the defects liability clause. There never was the necessity, because the contractor's liability for frost damage follows naturally from its general obligations under the contract.

The contractor is liable to make good frost damage caused by frost which occurred before practical completion – in other words, when the contractor was in control of the building works and could have taken appropriate measures to prevent the damage by introducing heating or stopping vulnerable work. Any damage caused by frost after practical completion is at the employer's own expense. The difference is usually easy to spot on site. Frost damage occurring after practical completion is often due to faulty detailing or maintenance.

9.3.4 Procedure

The rectification period starts on the date given in the certificate of practical completion as that on which practical completion was achieved or, in the case of partial possession, when possession of the part took place. The architect is to fill in the length of period required in the contract particulars. If no period is inserted, the length will be six months. The period should be agreed before tender stage with the employer, who will ask for the architect's advice. Although six months is a common rectification period, there are really no good reasons why the period should not be extended to nine or, better, 12 months. Specialist work such as heating often needs a 12-month period to fully test the system through all the seasons of the year. It is possible that a contractor asked to tender for a contract including a 12-month general rectification period would increase its tender figure slightly, but lengthening the period does not increase its actual liability, only as has been seen the contractor's right to return to make good defects. The increase in tender price probably reflects the confusion with 'maintenance period' and the fact that the final certificate would be delayed somewhat. In fact, only 2½% of the contract sum should be outstanding after practical completion or the last of the practical completion certificates (if there are sections) (clause 4.9).

Clause 2.30 simply refers to defects etc. 'which appear'. Although the time limit is the end of the period, the wording of the clause seems to give the architect the power to notify the contractor of all the defects which appear, including those which are present at the date of practical completion. Any other interpretation would make nonsense of the contractor's obligations: *William Tomkinson & Sons Ltd* v. *The Parochial Church Council of St Michael* (1990).

The architect is to notify defects to the contractor not later than 14 days after the expiry of the period. This is normally done as soon as possible after the end of the period. The architect should have inspected just before the period expired. There is no set form for notifying the contractor, but it is advisable for the architect to send it a letter (Figure 9.5) enclosing a schedule of defects. The architect's power to require defects to be made good is not confined to the issue of the schedule. The wording of the clause makes it clear that notification can be made to the contractor at any time within the period.

The requirement is for the contractor to make good the defects that are notified. No particular time limit is set, but it must carry out its obligation within a reasonable time. What is a reasonable time will depend on the circumstances, including the number of defects, their type, and any special arrangement to be made with the employer for access.

Ideally, the contractor should return to site within a week or so after receiving the architect's list and bring sufficient labour to make good the defects within, say, a month. If the architect decides to exercise the right to require defects to be made good during the currency of the period, it is good practice to confine such requests to really urgent matters in order to be fair to all parties.

All defects which are the fault of the contractor are to be made good at the contractor's own cost. However, there is an important proviso. If the employer agrees, the architect may instruct the contractor in a letter (Figure 9.6) not to make good some or all of the defects, and an appropriate deduction must be made from the contract sum. The clause gives no guidance on the method of arriving at an appropriate deduction. The job is best left to the quantity surveyor. Although no provision is made for the contractor to agree the amount of any such deduction, obviously it must be based on the cost of making the defects good by the contractor itself: *William Tomkinson & Sons Ltd* v. *The Parochial Church Council of St Michael* (1990). Although the architect does not have to give the contractor a reason for not requiring it to make good, such instructions will probably be issued only if the employer prefers to live with the defects rather than suffer the inconvenience of the contractor returning, or if the contractor's work record is so bad during the carrying out of the contract that the architect has no confidence that it will make a satisfactory job of making good. In the second instance, the architect

Figure 9.5
Architect to contractor, enclosing schedule of defects

Dear

The rectification period ended on [*insert date*]. I inspected the work on [*insert date*]. In accordance with clause 2.30 of the conditions of contract, I enclose a schedule of defects. I should be pleased if you would give them your immediate attention.

Yours faithfully

Copy: Clerk of works

Figure 9.6
Architect to contractor, instructing that defects are not to be made good

Dear

The rectification period ended on [*insert date*]. I inspected the Works on [*insert date*]. I enclose a schedule of the defects I found.

[*Then either:*]

I hereby instruct that, in accordance with clause 2.30, you are not required to make good any of the defects shown on the schedule.

[*Or:*]

I hereby instruct that, in accordance with clause 2.30, you are not required to make good those defects marked 'E'.

[*Then:*]

An appropriate deduction will be made from the contract sum in respect of the defects which you are not required to make good.

Yours faithfully

Copy: Employer
Quantity surveyor
Clerk of works

would probably be justified in obtaining competitive quotations for the work and deducting the amount from the contract sum. The employer would have to have a very good reason for preventing the original contractor from returning to make good trifling defects: *City Axis* v. *Daniel P. Jackson* (1998). The architect should obtain a letter from the employer authorising the instruction of the contractor that making good is not required (Figure 9.7).

When the architect is satisfied that the contractor has properly completed all making good of defects that have been notified, a certificate to that effect must be issued (clause 2.31). The certificate is important because it marks one of the dates starting the 28-day period within which the final certificate must be issued (clause 4.14).

The issue of the certificate of making good is much misunderstood. It is expressly stated to refer to the contractor's obligations under clause 2.30; in other words, the contractor's obligation to make good the defects notified under clause 2.30. The issue of the certificate cannot be lawfully withheld because further defects have come to light after the end of the rectification period. There is no doubt that the contractor is liable for such defects, but they are not defects to which the certificate is intended to relate. Ultimately, such defects may have to be dealt with by means of a deduction by the employer from the amount certified in the final certificate, subject to the issue of the relevant notices under clauses 4.14.2 and 4.14.3.

9.4 *Summary*

Possession

- Possession must be given on the date shown in the contract particulars; but
- It may be deferred for a maximum period stated in the contract particulars, not exceeding six weeks
- Deferment gives grounds for extension of time and claim for loss and/or expense
- Possession ends at practical completion.

Practical completion

- Practical completion requires the architect's certificate
- It is a matter for the architect's opinion alone subject to case law
- It marks the date on which many of the contractor's obligations end.

Figure 9.7
Architect to employer if some defects are not to be made good by the contractor

Dear

Further to our conversation/your letter/my letter [*delete as appropriate*] of [*insert date*], I enclose a list of defects found at the end of the rectification period. I understand that you wish me to instruct the contractor not to make good those defects which are marked 'E'/ any of the defects [*delete as appropriate*]. An appropriate deduction will be made from the contract sum.

In order that I may issue the instruction in accordance with the contract, I should be pleased to have your written consent to the course of action outlined above.

Yours faithfully

Partial possession

- Partial possession and sectional completion are for entirely different purposes.

Rectification period

- The rectification period is for the benefit of the contractor
- The end of the period does not mark the end of the contractor's liability for defects
- Defects can only be workmanship or materials not in accordance with the contract, or damage caused by frost occurring before practical completion
- The length must be entered in the contract particulars
- All defects apparent during the period or on practical completion are covered
- Defects must be notified to the contractor during the period and not later than 14 days after the end
- Defects are to be made good at the contractor's own cost; or
- If the employer agrees, the architect may instruct the contractor not to make good defects and deduct an appropriate sum from the contract sum
- After the contractor has made good, the architect must issue a certificate to that effect.

CHAPTER TEN
CLAIMS

10.1 General

This chapter covers claims by the contractor for both extra time and extra money, although there is not necessarily any link between the two. The subject of claims is an emotive one, but the architect has a duty under the contract to deal with claims, and in doing so to hold the balance fairly between the employer and the contractor. Although employed by the employer, in ascertaining and settling monetary claims and in making extensions of time, the architect has an independent role.

Two points should be noted. The first is that the architect's powers under the contract are limited. He or she can settle only claims that the architect is authorised to deal with under the express terms of the contract. This means that the architect cannot deal with common-law claims or make *ex gratia* settlements. To do that would require the express authority of the employer. However, note that clause 4.18.5 effectively covers breaches of contract by the employer. Table 10.1 summarises the contract clauses which may give rise to claims which the architect is empowered to settle and those which are the employer's province.

The second point is that, if the architect rejects a valid claim under the contract, case law establishes that this is a breach of contract for which the employer is responsible at common law: *Croudace Ltd* v. *London Borough of Lambeth* (1986). Whether or not in principle the contractor has a right to bring an action against the architect for failing to administer the claims provisions properly, the architect has a professional and contractual duty to act fairly and impartially as between the employer and contractor: *London Borough of Merton* v. *Stanley Hugh Leach Ltd* (1985). These observations apply to claims for extra money under the contract provisions or claims for 'direct loss and/or expense', as the contract puts it.

There is an important point to note about extensions of time: failure by the architect to exercise duties properly as to the granting of extensions of time may result in the contract completion date

Table 10.1

IC and ICD clauses which may give rise to claims (including paragraphs in schedules)

(see key on page 199)

Clause	Event	Type	Usually dealt with by
2.3	Fees and charges	C	A
2.5	Deferment of possession Deferment exceeding time in the contract particulars	C C	A A
2.7	Work not forming part of the contract	C	A
2.8	Contract documents not provided by architect Divulging or improper use of rates	C Cl	A E
2.9	Architect fails to determine levels	C	A
2.10 2.11	Architect fails to provide information in accordance with information release schedule or is late in provision	C	A
2.12 2.13	Errors and inconsistencies	C	A
2.14	Wrong valuation	Cl	E
2.15	Divergence from statutory requirements	C	A
2.16	Emergency compliance	C	A
2.17	Employer damage to materials on site	Cl	E
2.19	Architect's failure to give extensions of time	Cl	E
2.21	Failure to issue practical completion certificate	Cl	E
2.23	Improper deduction of liquidated damages	Cl	E
2.24	Failure to repay liquidated damages	Cl	E
2.25	Employer attempting to take possession without consent	Cl	E
2.30	Including items which are not defects	Cl	E
2.31	Failure to issue certificate of making good	Cl	E
3.3	Clerk of works exceeding powers	Cl	E
3.4	Employer failing to replace or replacing with unsuitable person	Cl	E
3.5	Architect unreasonably withholding consent to sub-letting	Cl	E

Table 10.1 *Cont'd*

Clause	Event	Type	Usually dealt with by
3.8	Wrongly phrased instructions	Cl	E
3.12	Postponement of work	C	A
3.14	Opening up and testing	C	A
3.15.2	Unreasonable instructions following failure of work	C	Adj or Arb
3.17	Unreasonable exclusion of person from site	Cl	E
4.3	Recovery of VAT	Cl	E
4.5	Failure to make advance payment	Cl	E
4.6	Certificate not issued or not issued at the proper time	Cl	E
4.8	Failure to serve notices	Cl	E
4.10	Retention money, interest	Cl	E
4.11	Following suspension by contractor	C	A
4.13.1	QS's failure to send computations to the contractor forthwith	Cl	E
4.14.1	Failure to issue final certificate, or certificate not in proper form	Cl	E
4.15, 4.16	Fluctuations	C	A
4.17	Disturbance of regular progress	C	A
5.2	valuation of variations not carried out in accordance with the contract	Cl	E
7.1	Assignment without consent	Cl	E
8.4	Invalid termination	Cl	E
8.5, 8.6	Invalid termination	Cl	E
Schedule 1 para A.4.4	Failure to pay insurance monies	Cl	E
Schedule 1 para C.4.4	Wrongful termination	C	Adj or Arb
Schedule 2 para 7.2	Contractor instructed to make own arrangements	C	A

Key: C = Contractual; Cl = Common law; E = Employer; A = Architect; Adj = Adjudicator; Arb = Arbitrator becoming unenforceable, in the sense that the contractor will no longer be bound by the contract period, but will merely be required to complete 'within a reasonable time'. The important consequence of this is that the employer would forfeit any right to recover liquidated damages, and would face the difficult and unenviable task of proving his actual loss at common law: *Percy Bilton Ltd* v. *Greater London Council* (1982). In such circumstances the architect might well be held to be negligent.

10.2 Extension of time

10.2.1 Legal principles

Under the general law, the contractor is bound to complete the Works by the agreed date, unless it is prevented from so doing by the employer's fault – and the employer's liability extends to the wrongful acts or defaults within the scope of the architect's authority provided that the employer had knowledge of such acts. Unless there is an extension of time clause in a contract, neither the architect nor the employer has any power to extend the contract period.

Clause 2.19 deals with the architect's power to grant extensions of time and lays down the procedure that must be followed. It is closely linked with clause 2.23 – which provides for liquidated damages – and case law establishes that clauses in this form will be interpreted very strictly by the courts. In practice, this means that if delay is caused by something not covered by the clause – or if the architect fails to exercise duties under it properly and at the right time – the employer will lose his right to liquidated damages: *Peak Construction (Liverpool) Ltd* v. *McKinney Foundations Ltd* (1970).

10.2.2 Liquidated damages

Clause 2.3 provides for the contractor to pay the employer liquidated damages at the rate specified in the contract particulars if the contractor fails to complete on time, or the employer may deduct the sum. The amount of liquidated damages should have been calculated carefully at pre-tender stage. The figure must represent a genuine pre-estimate of the loss likely to be suffered by the employer if the contractor fails to complete on time, or a lesser sum. If the sum arrived at is a genuine re-estimate of the likely loss, then that is the sum which will be recoverable, even if the sum agreed is greater than the actual loss or in the event there is no loss – the courts now take a pragmatic view: *Alfred McAlpine Capital Projects Ltd* v. *Tilebox Ltd* (2005). But liquidated damages are exhaustive of the employer's remedy for the breach of late completion: *Temloc Ltd* v. *Errill Properties Ltd* (1987). The calculation is sometimes difficult. In the case of profit-earning assets, there is no problem, and all the architect needs to do is analyse the likely losses and additional costs. The following should be considered:

- Loss of profit on a new building, e.g. rental income, retail profit
- Additional supervision and administrative costs
- Any other financial results of the contract's being late, e.g. staff costs
- On-costs under later direct contracts, e.g. a contract for fitting out
- Interest payable during the delay.

These points are not exhaustive, and clearly much depends on the type of project. For example, in the public sector late completion may require the temporary occupation of a more expensive building, and invariably there will be extra administrative costs and almost inevitably some financial penalties for late completion.

There are three possible bases for formulating liquidated damages in such cases:

- The notional rental value of the property can be taken, and, based on this, some notional return on its capitalised value can be calculated. This option is not recommended, nor its variant of fixing a sum commensurate with an average commercial project of comparable value
- The sources of likely loss can be used as a basis, for instance subsidies, capitalised interest on money advanced, and the spin-off effect on departmental costs. This method normally produces a nominal figure which does not encourage the tardy contractor to make up lost time. If the figure is unrealistic and the contractor has two contracts running late, it will be tempted to throw its resources into the contract which has the more realistic provision for liquidated damages
- Another method is to use formula calculation that gives an approximation of all individual costs. The best known is that put forward by the Society of Chief Quantity Surveyors in Local Government. This suggests three main headings under which a calculation should be made:

- Assume that 80% of the total capital cost of the scheme (including fees) will have been advanced at anticipated completion and that interest is being paid at the current rate. If that interest rate is 12%, capitalised interest is 80% × 12% ÷ 52 = 0.185% of the contract sum per week
- Assess administrative costs (e.g. staff salaries) as 2.75% of the contract sum per year. This gives 0.053% per week
- Exceptional costs (e.g. temporary accommodation) are assessed realistically.

The result of the first two parts of the formula reduces to 0.237% of the contract sum per week. This may be an underestimate of the likely losses, but it works reasonably in practice.

The main point about the calculation of liquidated damages is that the figure should not be plucked out of the air. A genuine attempt to calculate the likely loss must be made, and the resultant figure is used. In commercial projects the likely loss may produce a figure out of all proportion to the value of the contract. The solution then is to reduce the figure to an acceptable level.

Figure 10.1 is a possible format for a general calculation for liquidated damages. There are two preconditions for the deduction of liquidated damages under clause 2.23: *J.J. Finnegan Ltd* v. *Community Housing Association Ltd* (1995).

The first is the issue of a certificate of non-completion by the architect, which states that the contractor has failed to complete the Works or a section by the original or extended date for completion. This certificate is a factual statement, and more than one certificate can be issued. If an extension of time is made by the architect after issuing a certificate of non-completion, the contract provides that the extension will cancel the certificate, and the architect must issue a further certificate as necessary. Figure 10.2 is a pro-forma example.

The second is that, before the date of the final certificate, the employer must inform the contractor that the employer may require the contractor to pay or deduct liquidated damages. This must be done in writing to the contractor to that effect. If, unusually, payment is required, it ranks as a debt that the employer may recover in the usual way. This notice is in addition to the notices of withholding payment which must be given under clauses 4.8.3 or 4.14.3 depending on whether the payment is to be deducted from interim payment or after the issue of the final certificate.

If the certificate of non-completion is cancelled because the architect has extended time, any liquidated damages deducted by the employer in the meanwhile must be repaid to the contractor

Figure 10.1
Calculation of liquidated and ascertained damages (typical format)

Contract ....

Client ....

Architect ....

		costs/week
1. SUPERVISORY STAFF (current rates)		
Architect:		
Estimated hrs/wk . . . × time charge of £ . . . /hr		£
Quantity surveyor:		
Estimated hrs/wk . . . × time charge of £ . . . /hr		£
Consultants [*as above for each one*]		£
Clerk of works		
Weekly salary (= yearly ÷ 52)		£
	Total (1)	£
2. ADDITIONAL COSTS (current rates)		
Rent and/or taxes and or charges for present premises		£
Rent and/or taxes and or charges for alternative premises		£
Charges for equipment		£
Movement of equipment		£
Additional and/or continuing and/or substitute staff		£
Movement of staff (include travel expenses)		£
Other travel and parking		£
Any site charges which are the responsibility of the client		£
Extra payments to directly employed trades		£
Insurance		£
Additional administrative costs		£
	Total (2)	£

Figure 10.1 *Cont'd*

		costs/week
3. INTEREST		
Interest payable on estimated capital expenditure up to the contract completion date, but from which no benefit is derived. Estimated expenditure taken as 80% of contract sum and fees.		
Contract sum		£
Architect's fees (90%)*		£
Quantity surveyor's fees (90%)**		£
Consultants' fees (90%)**		£
Salary of clerk of works (£/wk × contract period)		£
Interest charges at current rate of £ . . . %: interest therefore		
$= \dfrac{80\% \text{ capital expended} \times \text{interest}}{52}$		£
	Total (3)	£
4. INFLATION		
Current rate of inflation . . . %/year		
Total (1) × . . . % × contract period (in years)		£
Total (2) × . . . % × contract period (in years)		£
	Total (4)	£
5. TOTAL LIQUIDATED AND ASCERTAINED DAMAGES/WEEK		
	Total (1)	£
	Total (2)	£
	Total (3)	£
	Total (4)	£
	Final total:	£

* It is essential that all costs are additional, i.e. they would not be incurred if the contract was completed on the contract completion date.
The headings are given as examples. Every project is different.

** Professional fees are taken as 90% of total, because some professional work remains to be done after practical completion.

Figure 10.2
Certificate of non-completion under clause 2.22

To the employer

CERTIFICATE OF NON-COMPLETION

In accordance with clause 2.22 of the above contract, I hereby certify that the contractor [*insert name*] has failed to complete the Works by the date for completion stated in the contract particulars/within the extended time fixed by me under clause 2.19 [*delete as appropriate*], namely [*insert date*].

Dated this............ day of....................... 20......

[Signed] .... Architect

Copy: Contractor

(clause 2.24). For example, the architect has issued a certificate of non-completion and the employer has deducted four weeks' liquidated damages at £750 a week, making a total of £3000. The architect then issues a variation instruction and assesses the resultant delay at three weeks, thereby cancelling the first certificate and issuing a new one once the extended date is passed. The employer must then repay 3 × £750 = £2250 to the contractor. The employer repays this sum net; the contractor has no right to interest on the money, either under the contract or under the general law.

10.2.3 Procedure

Clauses 2.19 and 2.20 are basically a simplified version of SBC, clauses 2.26 to 2.29, but the provisions for the contractor to notify the architect of delays affecting progress are less detailed, and there are other important differences. The flowchart in Figure 10.3 illustrates the contractor's duties in claiming an extension of time under clause 2.19. The flowchart in Figure 10.4 sets out the duties of the architect in relation to such a claim. The case of *Royal Brompton Hospital NHS Trust* v. *Hammond and Others (No. 7)* (2001) contains much useful advice for the architect when making an extension of time. The procedure under clause 2.19 is as follows.

As soon as progress is, or is likely to be, delayed – the contract says 'If and whenever it becomes reasonably apparent' – the contractor must give the architect written notice of the cause of the delay forthwith. The wording of clause 2.19.1 imposes a duty on the contractor to notify the architect of any delay (or likely delay) to progress, and it is not confined to notifying the architect of the events listed later.

No particular form of notice is prescribed, but if the notification does not give sufficient detail, the architect can require the contractor to provide such further information as is reasonably necessary to enable the discharge of functions under the clause (2.19.4.2). Figure 10.5 is a suitable letter from the contractor to the architect notifying a delay giving grounds for extension of time, and Figure 10.6 is an example of a letter providing further information.

On receipt of the notice and any necessary supporting information, if the architect thinks that the completion of the Works or a section is being, or is likely to be, delayed beyond the particular completion date by one or more of the relevant events specified in clause 2.20, a 'fair and reasonable' extension of time for completion must be given to the contractor. This must be done as soon as the architect is able to estimate the length of the delay. In making this estimate, the architect should take into account the overall proviso in clause 2.19.4.1. This proviso is in two parts: the contractor is required constantly to use its best endeavours to prevent (not to reduce) delay (e.g. by reprogramming); and it must do all that may be reasonably required to the satisfaction of the architect to proceed with the Works. What steps the contractor must take depends on the circumstances of the case, but the architect cannot require it to accelerate progress, and its obligation stops short of expending substantial sums of money.

There is no specific time limit, but the architect must grant a fair and reasonable extension of time for completion as soon as the length of the delay can be estimated beyond the currently fixed completion date. The various decided cases suggest that it must be done as soon as reasonably possible, although where there are multiple causes of delay there may be no alternative but to leave the decision until a later stage: *Amalgamated Building Contractors Ltd* v. *Waltham Holy Cross UDC* (1952). Figure 10.7 is a suggested letter from an architect to a contractor awarding an extension.

Indeed, because the architect has a 12-week period from practical completion in which to grant further extensions of time if the circumstances warrant it (clause 2.19.3), it may be argued that parsimony is the order of the day – contractors will certainly take a different view! Failure to grant an extension of time properly may result in the contract completion date becoming at large, so the architect must be careful, although there is some room to manoeuvre.

The contractor's written notice is not in fact a precondition (or 'condition precedent' in lawyer's language) to the grant of an extension of time, because the clause goes on to empower the architect to grant an extension of time up to 12 weeks after the date of practical completion 'whether upon reviewing a previous decision or otherwise and whether or not the Contractor has given notice'. The architect must therefore consider the contract as a whole and decide whether any extension of time is justified even if the contractor has failed to give proper or any notice of delays to progress. Unless the architect does this within the time limit laid down, the employer's right to liquidated damages will be jeopardised. It is sometimes said that the case of *Temloc Ltd* v. *Errill Properties Ltd* (1987) is authority that the 12-week period is not really binding. It is, of course, authority for no such thing, and the Court of Appeal's view on that point depended on the particular facts of that case which involved the employer attempting to take advantage of the architect's breach of contract.

When considering extensions of time after practical completion, the architect cannot reduce any extension of time previously granted.

Figure 10.3
Contractor's duties of there is a delay (clause 2.19)

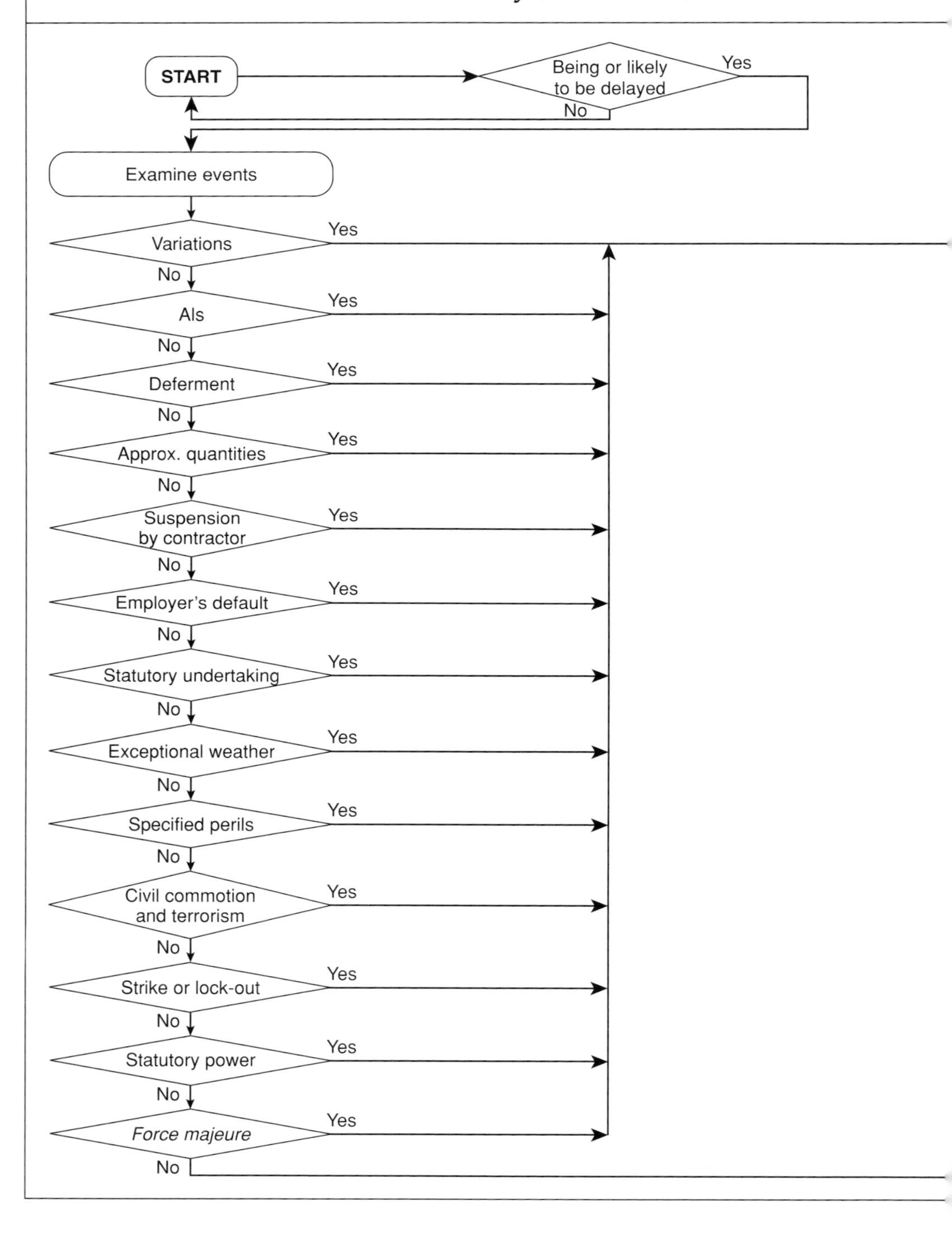

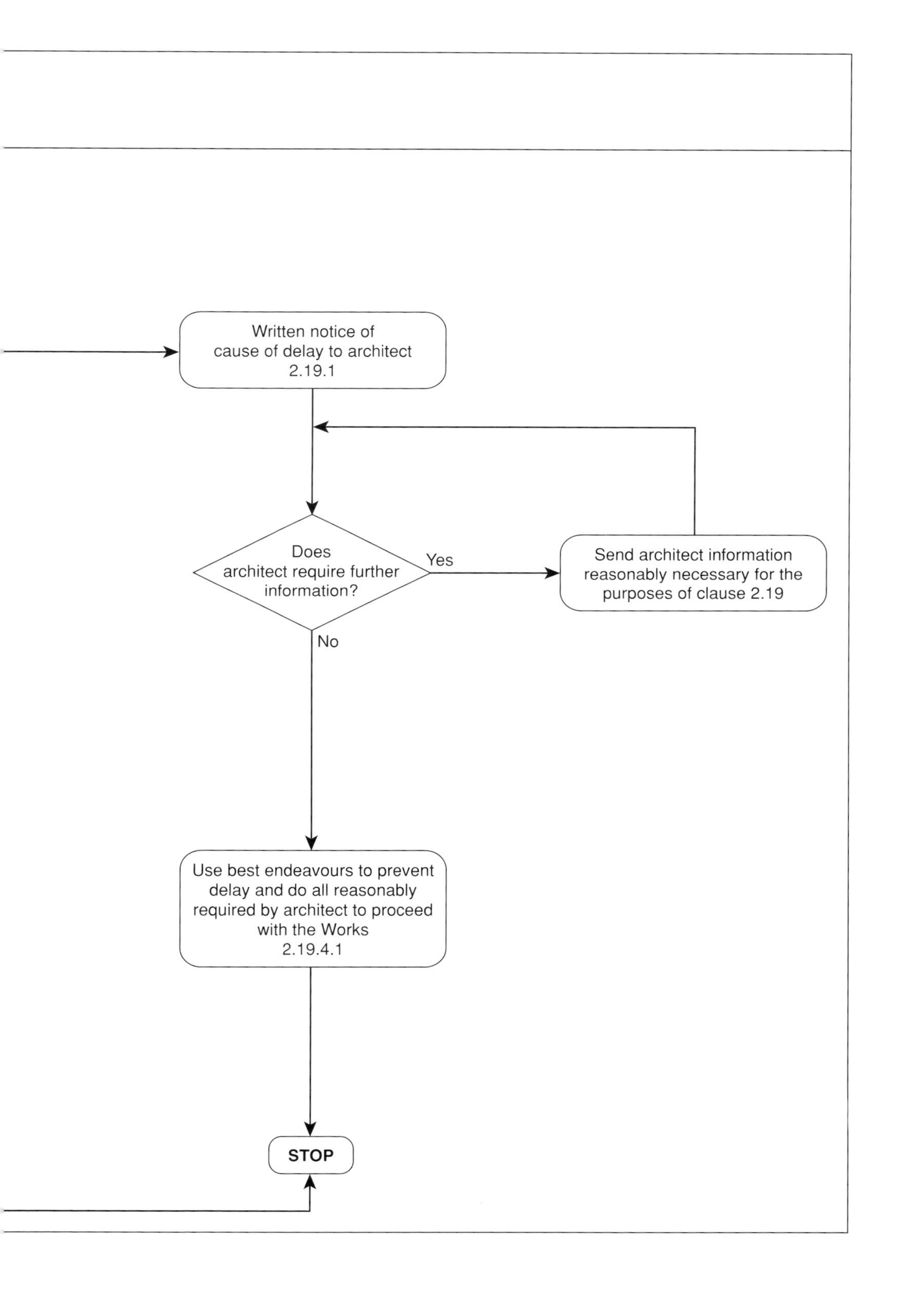
Written notice of
cause of delay to architect
2.19.1
Does
architect require further
information?
Yes
Send architect information
reasonably necessary for the
purposes of clause 2.19
No
Use best endeavours to prevent
delay and do all reasonably
required by architect to proceed
with the Works
2.19.4.1
STOP

Figure 10.4
Architect's duties in relation to extension of time (clause 2.19)

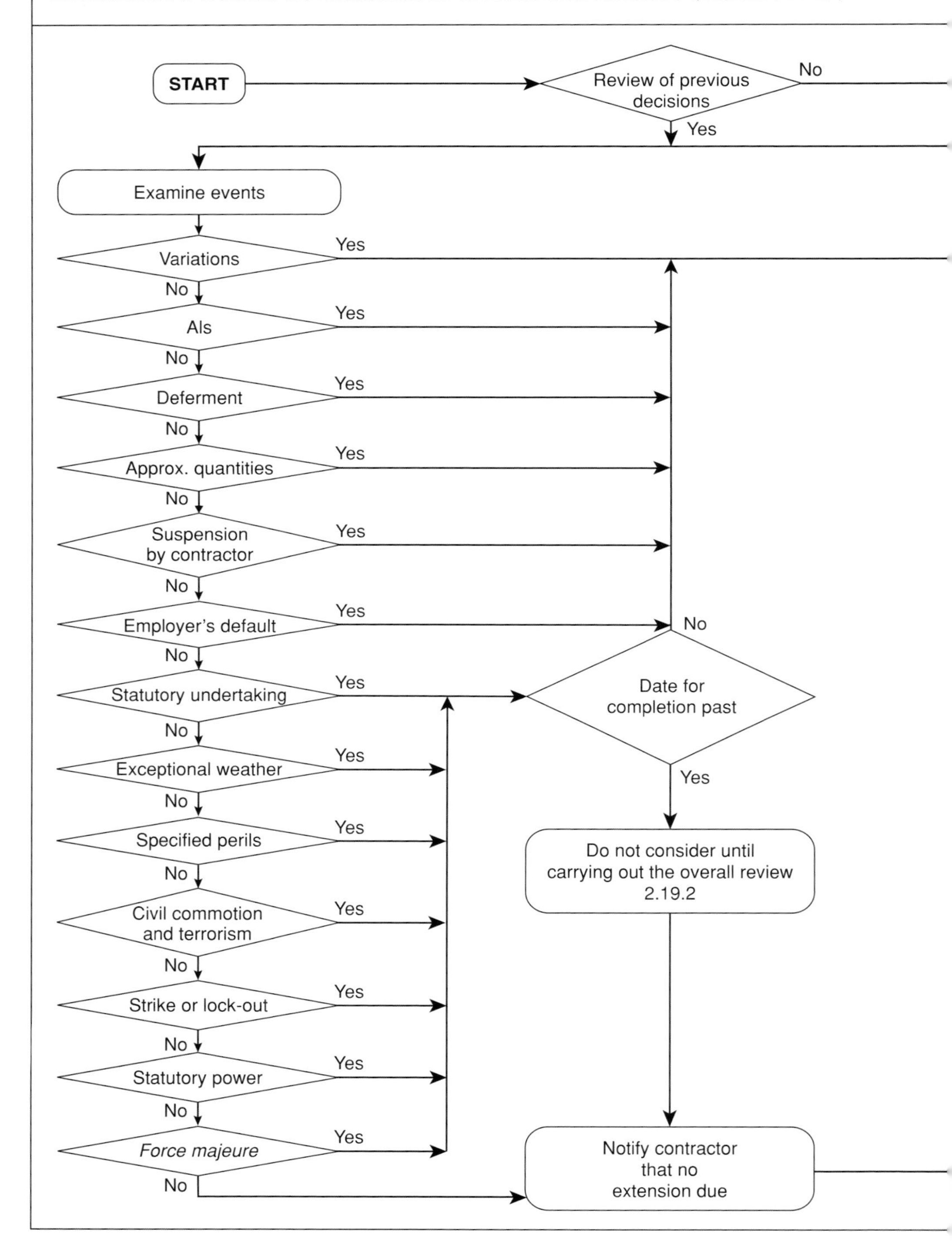

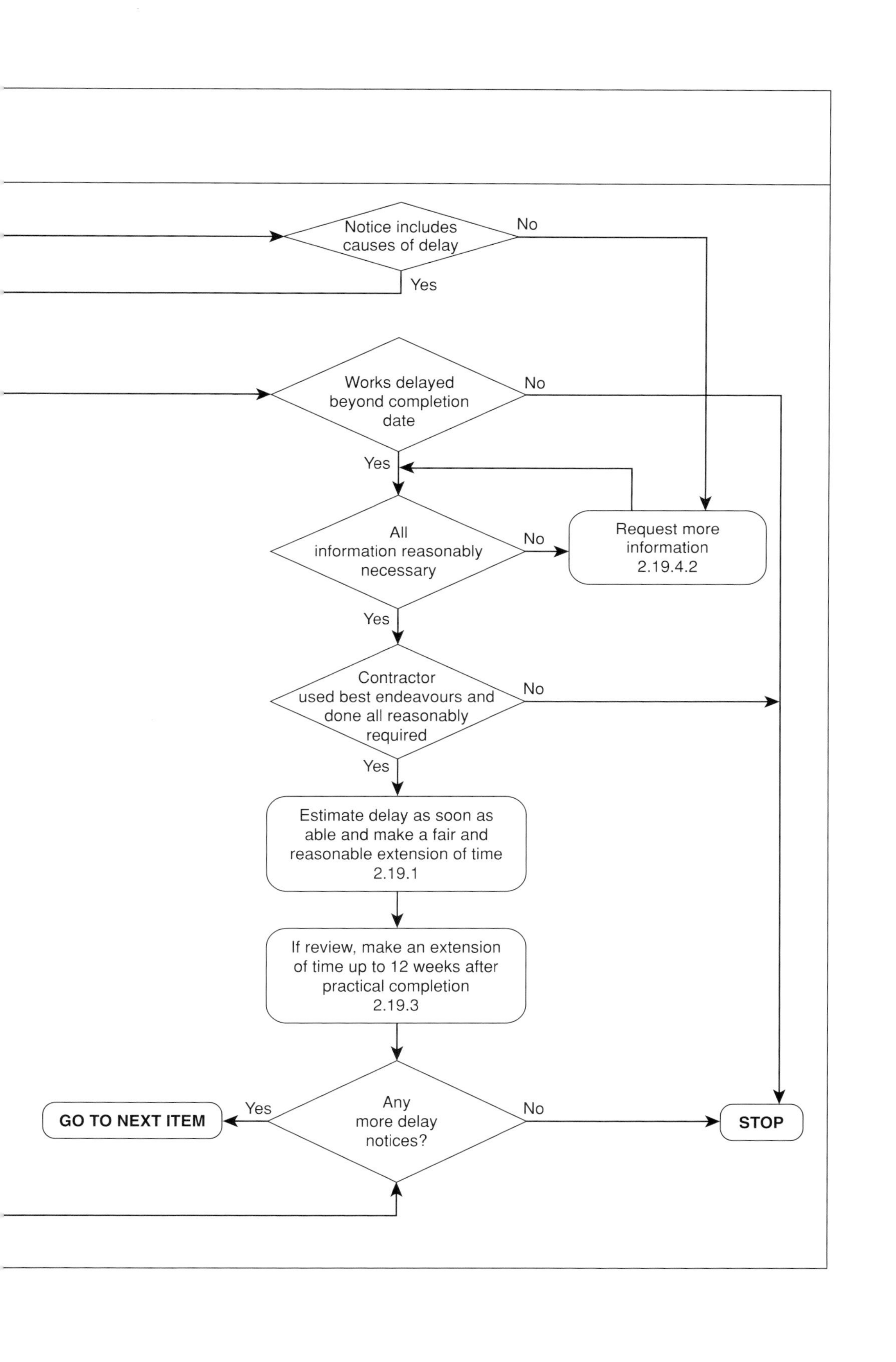

Notice includes causes of delay
No
Yes
Works delayed beyond completion date
No
Yes
All information reasonably necessary
No
Request more information 2.19.4.2
Yes
Contractor used best endeavours and done all reasonably required
No
Yes
Estimate delay as soon as able and make a fair and reasonable extension of time 2.19.1
If review, make an extension of time up to 12 weeks after practical completion 2.19.3
GO TO NEXT ITEM
Yes
Any more delay notices?
No
STOP

Figure 10.5
Contractor to architect, notifying delay

Dear

In accordance with clause 2.19.1 of the conditions of contract, we give you formal notice that the progress of the Works is being delayed by [*specify event*], which is a ground for the grant of an extension of time under clause 2.19.

In order to assist you in reaching a decision, we provide the following information [*specify*]. This, as you will see, is likely to cause further delay to [*specify activity*]. We are, of course, using our best endeavours to minimise the delay, but we must ask you to exercise the power of granting an extension of time for completion in accordance with clause 2.19.

Yours faithfully

Figure 10.6
Contractor to architect, providing further information

Dear

Thank you for your letter of [*insert date*] requesting further information in respect of [*state what the architect requires*].

[*Either:*]

As you request, we enclose [*specify enclosures*]. We trust that this information is sufficient for your purposes.

[*Or, if the architect is not specific in the request:*]

However, we note that clause 2.19.4.2 requires us to provide such information you require as is 'reasonably necessary' for the purposes of clause 2.19. It is not clear from your letter what further information you require. If you would be good enough to be specific, we shall be pleased to provide you with the information to the extent that it is reasonably necessary.

Yours faithfully

Figure 10.7
Architect to contractor, making extension of time under clause 2.3

Dear

I refer to your notice of delay of [*insert date*]. [*If appropriate, add:*] and the further information provided in your letter of [*insert date*].

In accordance with clause 2.19.1 of the conditions of contract, I hereby make an extension of time of [*specify period*], the revised date for completion now being [*insert date*].

Yours faithfully

The effect of the contractor failing to notify the architect of delays as the clause requires is that it loses the benefit of an extension of time during the currency of the contract, but it is still entitled to an extension after practical completion if the circumstances warrant it. This is especially important if delay has been caused by one of the specified events which are the employer's responsibility in law (e.g. late instructions or details). Therefore it is suggested that, as a matter of practice, the architect should review every contract and write to the contractor accordingly, even though the contract does not require that to be done. Figure 10.8 is a suitable letter for the architect to send to the contractor.

However, failure by the contractor to notify delay at the correct time is a clear breach of contract, which the architect is entitled to take into account when making an extension of time subsequently: *London Borough of Merton* v. *Stanley Hugh Leach Ltd* (1985). This will most likely take the form of a reduction in the extension of time that the contractor might otherwise expect if, viewed objectively, the failure to notify prevented the architect from taking steps to reduce the delay.

Although the provisions for the granting of extensions of time are explicit, considerable problems arise in practice. It cannot be overemphasised that clauses 2.19 and 2.20 are for the benefit of the employer as well as of the contractor. The proper exercise by the architect of clause 2.19 prevents the contract from becoming 'at large' and so preserves the employer's right to liquidated damages. The granting of an extension of time relieves the contractor of the liability to pay liquidated damages from the specified date and allows extra time in which to complete the Works where delay has been caused by events which the contract draftsman has put at the employer's risk.

Many of the problems arise because the architect and the contractor are imperfectly aware of their obligations. Failure by the contractor to give notice of delay at the right time and to provide supporting information often results in extensions being granted later rather than sooner. It has to be admitted, however, that the fault is sometimes on the architect's side, and there have been cases where (unwisely) the architect has made unreasonable requests for further information in order to postpone the granting of an extension of time. Figure 10.9 is a letter which may be useful in such circumstances. If the architect delays unreasonably in granting an extension of time, the contractor should write a letter along the lines of Figure 10.10.

Figure 10.8
Architect to contractor, after reviewing extensions of time

Dear

In accordance with clause 2.19.3 of the conditions of contract,

[*Either:*]

I confirm the date of completion as being [*insert date*]

[*Or:*]

I confirm the date of completion as extended as being [*insert date*]

[*Or:*]

I hereby make an extension of time of [*specify period*] to take account of [*specify*], thus giving a revised date for completion of [*insert date*].

Yours faithfully

Figure 10.9
Contractor to architect who unreasonably requests further information

Dear

Thank you for your letter of [*insert date*] in which you request further information in respect of [*specify*] to enable you to make an extension of time.

On [*insert date*] we notified you of a delay to progress as required by the conditions of contract, and with that notice we enclosed full details of the effects of the notified event on progress by reference to our programme. This showed that the delay was on the critical path. In response to your letter of [*insert date*] we provided you with the further information you requested.

We believe that we have given you the fullest information to enable you to reach a decision. Clause 2.19.4.2 limits our obligation to providing such information you require as is 'reasonably necessary' for the purpose of clause 19. We regret that we consider your reply to be an attempt to postpone the making of an extension of time. This is in neither the interests of the employer nor of ourselves, and we must call on you to exercise the duty imposed on you by clause 2.19.1 and to make a fair and reasonable extension of time for completion.

Yours faithfully

Figure 10.10
Contractor to tardy architect

Dear

On [*insert date*] we gave you formal notice of delay under clause 2.19.1 of the conditions of contract and requested you to make a fair and reasonable extension of time for completion. At your request on [*insert date*] we provided you with further information to enable you to reach a decision, namely [*specify information provided*].

It is now [*insert number*] weeks since the date of our notice and we consider that you have had more than sufficient time in which to reach your decision, particularly as you are aware that, because of the delay, the [*specify activity*] work has had to be postponed. We shall be very grateful if you would proceed at once to make the necessary extension of time in the interests of both the employer and ourselves.

Yours faithfully

10.2.4 Grounds

IFC 98 had 18 relevant events. The new contracts have rationalised the content, combining some and omitting others which were already covered. There are now 13 events which may give rise to an extension of time specified in clause 2.20, and the architect can extend the time for completion only if one or more of these events occurs and causes, or is likely to cause, delay to progress. The events listed are traditional, and they largely parallel those in SBC.

The events divide into two distinct and separate groups: those which may also, quite independently, found a claim for loss and/or expense if application is properly made under clauses 4.17 and 4.18; and those which merely entitle the contractor to an extension of time.

The first group consists of events which are the responsibility of the employer, either personally or through those for whom there is vicarious responsibility in law. Events giving rise to time and, independently, to direct loss and/or expense are as follows.

Variations

This refers to variations which arise from instructions or from anything which the contract provides is to be treated as a variation (clause 2.20.1).

Architect's instructions

Those referred to are: clause 2.13, inconsistencies; clause 3.12, postponement of any work to be executed under the contract; clause 3.13, expenditure of provisional sums; clause 3.7, sub-contractors to the extent specified in clause 3.7 itself and schedule 2; clauses 3.14 and 3.15.1, opening up or testing and failure of work – to the extent that the contractor is not at fault (clause 2.20.2).

Deferment of possession

This will give rise to an extension of time when the contract particulars state that clause 2.5 is to apply and an extension of time is an almost necessary concomitant of deferment of possession (2.20.3).

Approximate quantity not a reasonably accurate forecast

This applies where approximate quantities are given and they are not reasonably accurate. If they forecast a smaller quantity than the quantity actually required, the contractor is entitled to an extension for the actual extra time taken to carry out the work (2.20.4).

Contractor's suspension of obligations

This ground applies where the contractor has properly suspended its obligations under the contract clause 4.11 because the employer has failed to pay in full by the final date for payment (2.20.5).

Employer prevention or default

This is designed to mop up any action or inaction of the employer which could give rise to delay. Some individual items were listed in IFC 98, such as failure to give site access and delays caused by directly employed operatives. The clause covers acts by the architect, the quantity surveyor and any person for whom the employer is responsible, but not so far as any act of the contractor contributed to it (clause 2.20.6).

The important point about all the following events is that delay from one of them does not give rise to any claim for extra cost under clause 4.17 but merely to a claim for extension of time. Their common characteristic is that they are outside the control of either contracting party.

Local authority or statutory undertaker's work

This covers only work done 'in pursuance of . . . statutory obligations in relation to the Works' or failure to carry out such work. If a statutory undertaker (e.g. a gas supplier) does work under contract with the employer, this would fall under clause 2.7 and be dealt with under clause 2.20.6 as regards extension of time (clause 2.20.7).

Exceptionally adverse weather conditions

The key word is 'exceptionally', because the contractor is expected to programme to take account of the sort of weather normally to be expected in the area at the relevant time of year. Moreover, the mere occurrence of exceptionally adverse weather – such as a fall of snow in the Midlands in mid-June – is not sufficient. It must actually delay progress: *Walter Lawrence* v. *Commercial Union Properties* (1984) (clause 2.20.8).

Specified perils

These are the usual insurance risks – fire, tempest, and so on (clause 2.20.9).

Civil commotion or use or threat of terrorism

The essential element of civil commotion is turbulence or tumult. Public protest gatherings may sometimes fall into this category. The threat would have to be substantial and not just a general fear because incidents had occurred elsewhere (clause 2.20.10).

Strikes and similar events

For the most part, the list of events is self-evident, and the breadth of the strike clause should be noted. It covers strikes, etc. affecting any of the trades employed on the Works and also those engaged in preparing or transporting materials needed for the Works (clause 2.20.11).

Exercise of statutory powers

This ground must be a direct edict from the government, which must affect the execution of the Works directly. It is thought that the closure of some access roads during the foot and mouth epidemic probably falls under this head insofar as such roads were essential to access the Works (clause 2.20.12).

Force majeure

Although wider in its meaning than the English term 'Act of God', this has a restricted meaning here, because some matters which would otherwise constitute *force majeure* are dealt with expressly. For an event to amount to *force majeure*, it must be catastrophic and outside the control or contemplation of either contracting party (clause 2.20.13).

10.3 Loss and expense claims

10.3.1 Definition

Clause 4.17 gives the contractor a right to be reimbursed 'direct loss and/or expense' which it incurs as the result of the events specified and for which there is no other payment under the contract, provided the contractor follows the procedure laid down. Claims for loss and expenses are a regulated provision for the payment of sums equivalent to damages.

'Direct loss and/or expense' is to be equated with the damages recoverable for breach of contract at common law: *F.G. Minter Ltd* v. *Welsh Health Technical Services Organisation* (1980). The purpose of the clause is to set out the contractor's rights in anticipation of the specified relevant matters, and is the means of putting the contractor back into the position in which it would have been but for the delay or disruption.

As the settlement amounts to damages at common law, an exact establishment of the contractor's additional costs must be made. Many contractors hold a fallacious view about claims generally, and the following three important points are often overlooked:

- The contractor must 'mitigate its loss', i.e. take reasonable steps to diminish the loss, for example by redeploying resources
- As damages are subject to the common-law 'foreseeability test', the contractor can recover only that part of the resultant loss and/or expense that was reasonably foreseeable to result. This is to be judged at the time the contract was made and not in the light of the events that have occurred
- The loss and/or expense must be direct. It must have been caused by the event relied on without any intervening cause: *Saint Line Ltd* v. *Richardsons Westgarth & Co Ltd* (1940).

10.3.2 Procedure

The contract requires the contractor to make a written application to the architect within a reasonable time of its becoming apparent that it has incurred, or is likely to incur, loss or expense resulting from specified causes. These are that the employer defers giving possession of the site under clause 2.5 (if applicable); or that the regular progress of the Works is 'materially' affected by one or more of the five relevant matters set out in clause 4.18. 'Materially' means to a substantial extent and not trivially.

The contractor's written application need not be in any particular set form, but the architect may require the provision of further information to enable the architect (or the quantity surveyor) to assess the claim. The sort of information that may be required will depend on the circumstances, and it is the contractor's duty to provide the architect (or the quantity surveyor) with sufficient documentary evidence to enable the claim to be properly considered. The contract refers to the information being 'reasonably necessary'. Figures 10.11 and 10.12 may be used by the contractor as pro-formas. When the contractor has made a written application, the architect must decide whether it has incurred, or is likely to incur, loss and/or expense resulting from one or more of the relevant matters, and the architect must also be satisfied that the contractor is not receiving payment under some other contract provision for the same thing (e.g. under clause 3.11 in respect of variations).

If the architect is so satisfied, then the amount of the loss of expense must be ascertained. This can be done by the architect or more usually, if so instructed, by the quantity surveyor. Figures cannot be plucked out of the air: 'ascertain' means to establish definitely, and is not the same as 'estimate' or 'guess' (*Alfred McAlpine Homes North Ltd* v. *Property and Land Contractors Ltd* (1995)), although some element of judgment may be involved: *How Engineering Services Ltd* v. *Lindner Ceilings Partitions plc* (1999). The amount so ascertained is to be included in the next payment certificate.

The contract allows additional or alternative claims for breach of contract based on the same facts, as is made clear by clause 4.19, which preserves the contractor's normal rights at common law. Because of this provision, the contractor may also bring common law claims, which are often based on implied terms relating to the employer's duty to co-operate and not to hinder. The architect has no power or authority to deal with common law claims: only claims arising under clause 4.17. From the contractor's point of view, the fact that its common law rights are preserved is a very real benefit, because if it has failed to make the written application required by clause 4.17, and so has lost the right to reimbursement under the contract terms, it can bring a claim at common law based on the same facts, provided that the event relied on is also a breach of contract (e.g. late instructions from the architect). Such claims, however, must be pursued in adjudication, arbitration or litigation.

The flowchart in Figure 10.13 sets out the contractor's duties under clauses 4.17 and 4.18. The flowchart in Figure 10.14 sets out the architect's duties under the provision.

10.3.3 Matters grounding a claim

Five 'matters' are listed in clause 4.18, and it is the occurrence of one of these – or the employer's deferment of giving possession of the

Figure 10.11
Contractor to architect, written application for reimbursement of direct loss and/or expense

Dear

In accordance with clause 4.17 of the conditions of contract, we hereby make written application to you for reimbursement of direct loss and/or expense in the execution of this contract for which we will not be reimbursed by payment under any other contractual provision, because regular progress of the Works has been materially affected by [*specify, e.g. compliance with your instruction number 3 or as appropriate*], this being a relevant matter specified in clause number [*insert number*].

Yours faithfully

Figure 10.12
Contractor's response to architect's request for further information

Dear

Thank you for your letter of [*insert date*] in which you request us to provide further information in support of our application for reimbursement of direct loss and/or expense dated [*insert date*].

We have pleasure in enclosing the following: [*Specify the information enclosed*].

We trust that this completes the information reasonably necessary for the purposes of clause 4.17.

Yours faithfully

Figure 10.13
Contractor's duties in applying for loss and/or expense (clause 4.17)

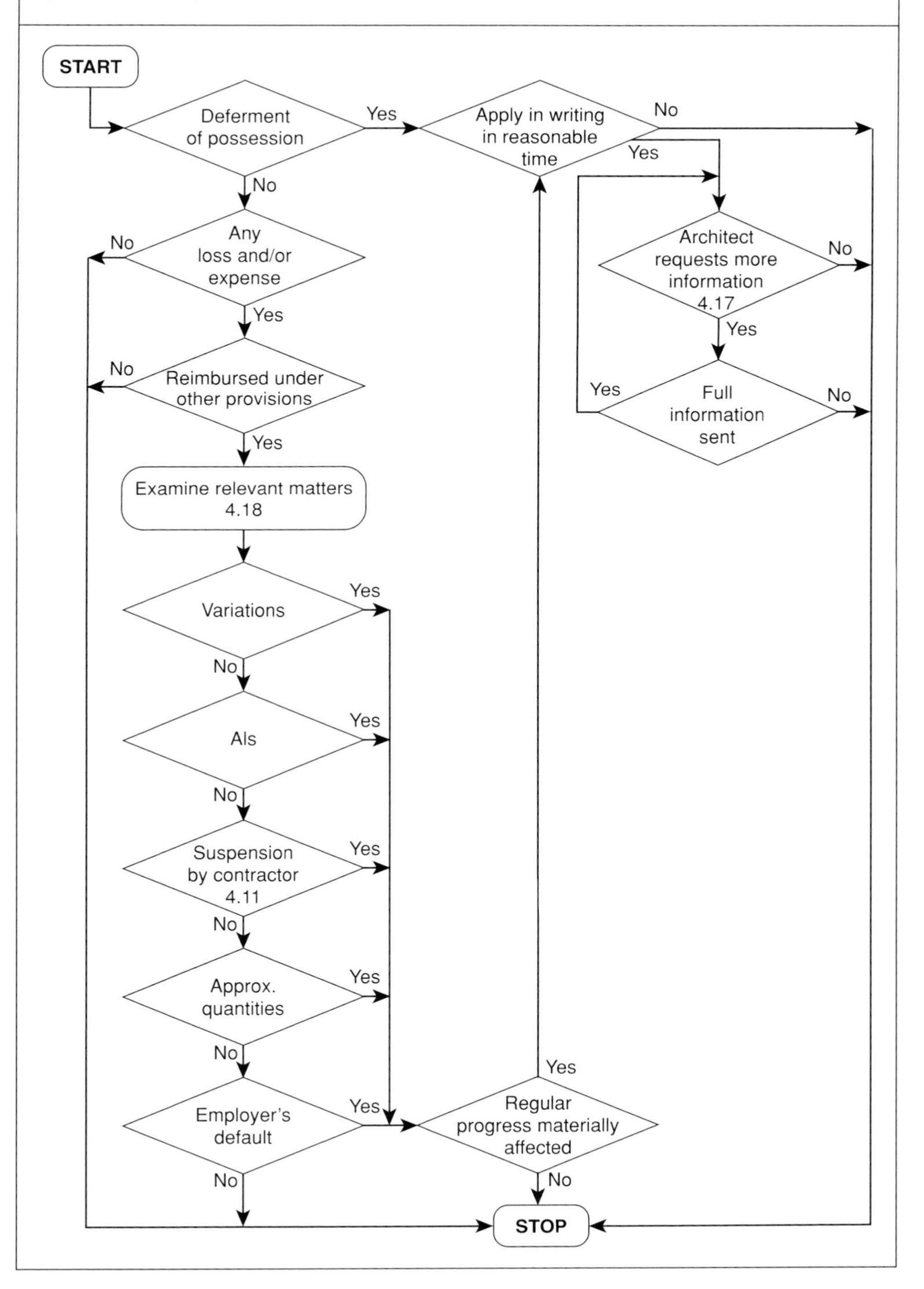

Figure 10.14
Architect's duties in relation to loss and/or expense application (clause 4.17)

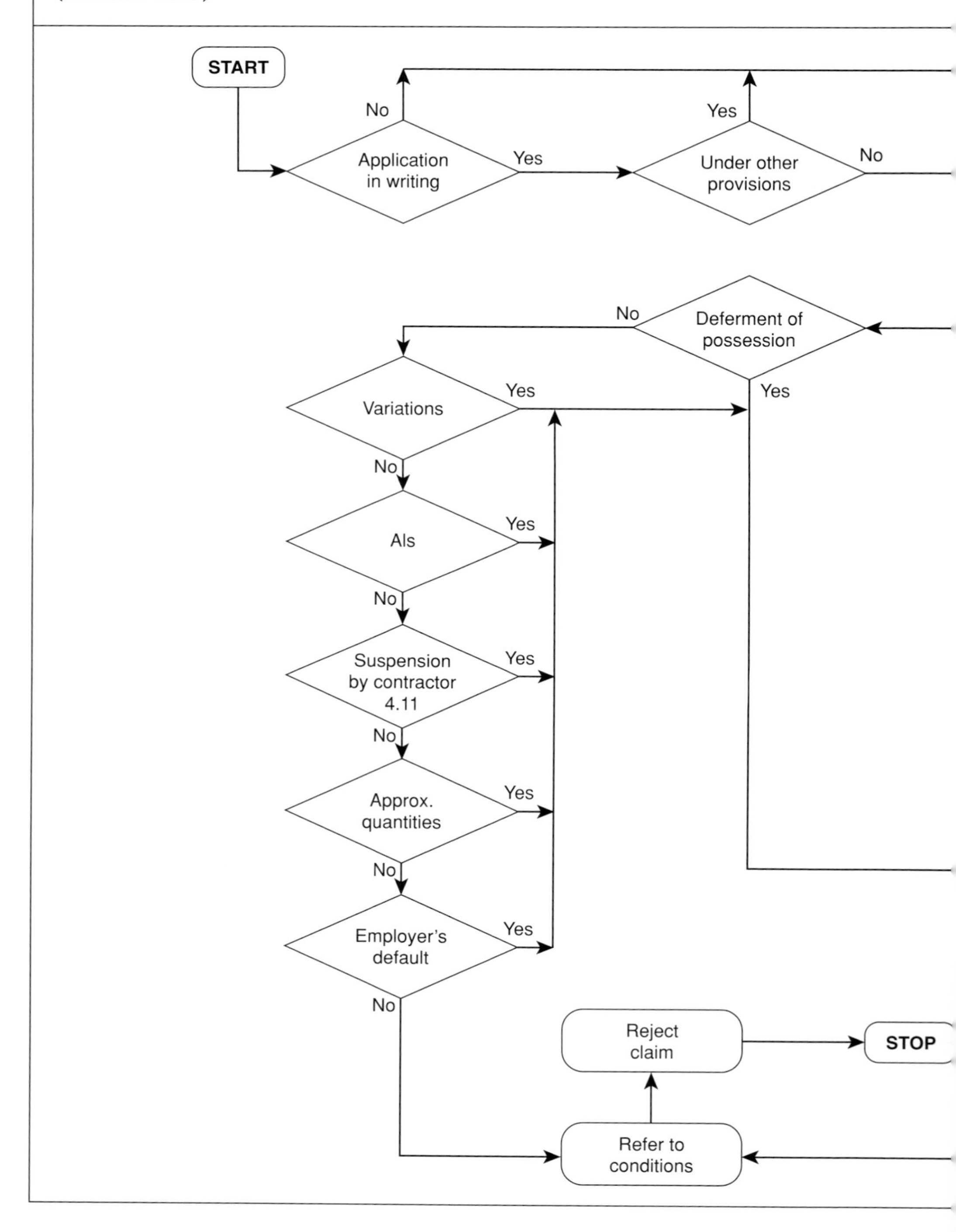

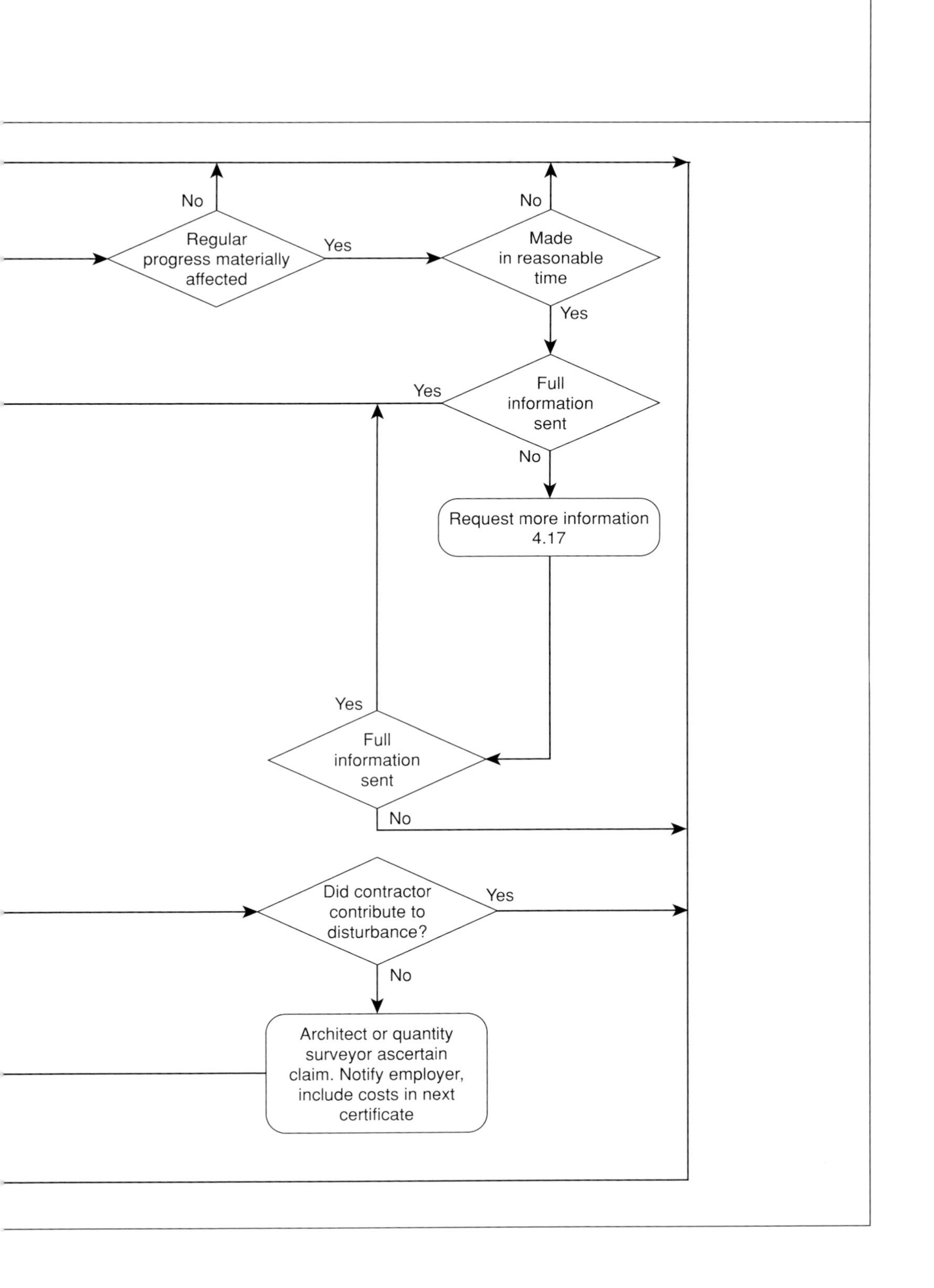
No
Regular progress materially affected
Yes
No
Made in reasonable time
Yes
Yes
Full information sent
No
Request more information 4.17
Yes
Full information sent
No
Did contractor contribute to disturbance?
Yes
No
Architect or quantity surveyor ascertain claim. Notify employer, include costs in next certificate

site – which triggers off a claim. The clause 4.18 relevant matters are as follows:

- Variations: arising from instructions or other matters
- Architect's instructions: those referred to are issued under clause 3.12 (postponement); 3.13 (provisional sums); 3.7 and schedule 2 (named sub-contractors, as specified in that clause); 3.14 (opening up and testing); 2.13 (inconsistencies)
- Suspension by the contractor of its obligations under clause 4.11
- Approximate quantity not a reasonably accurate forecast
- Employer prevention of default: except to the extent that the contractor has contributed to it.

10.4 Summary

Claims for time and money are distinct; there is no necessary connection between the two.

Liquidated damages

Liquidated damages are:

- A genuine pre-estimate of likely loss or a lesser sum
- Recoverable without proof of loss
- Recoverable by deduction under the contract only if the architect issued a certificate of delay and the employer has notified the contractor in writing of an intention to require liquidated damages.

Extension of time

The architect is bound to grant a fair and reasonable extension of time for completion on the happening of certain events. Failure to do so may result in the contract date becoming 'at large' and liquidated damages being irrecoverable.

The contractor must observe the contract notice procedure and, if it does so, the architect must:

- Grant in writing a fair and reasonable extension of time for completion as soon as practicable
- Review contract progress within 12 weeks of practical completion and adjust the contract time accordingly, even if the contractor has not notified him or her of an event giving rise to an extension of time

- Grant an extension of time only on grounds specified in the contract.

Loss and expense claims

The contractor has a right to be reimbursed 'direct loss and/or expense' under the contract on the happening of certain disruptive relevant matters provided it invokes the contract procedures. The contractor must:

- Make written application to the architect at the right time
- Provide the necessary supporting evidence.

The architect must:

- Be satisfied that the notified event is a valid ground of claim
- Ascertain (or instruct the quantity surveyor so to do) the amount of loss or expense directly incurred
- Include the sum so ascertained in the next certificate of payment.

The architect has no power under the contract to deal with common law or *ex gratia* claims.

CHAPTER ELEVEN
PAYMENT

11.1 The contract sum

The figure shown in article 2 of the articles of agreement is of great importance. It is the amount for which the contractor has agreed to carry out the whole of the Works. In contracts such as IC and ICD, which are known as 'lump sum contracts', the contractor is entitled to payment provided it substantially completes the whole of the Works. The fact that the contract provides for interim payments does not alter the position. If the contractor abandons the Works before they are finished, the employer is entitled to pay nothing more. The principle of interim payments is to provide sufficient money to allow the contractor to carry out the Works. It is a system to provide cash flow to the contractor. It is purely a business arrangement. It matters not that the contractor has imparted considerable benefit to the employer. If it does not substantially complete the Works, it is not entitled to payment: *Hoenig* v. *Isaacs* (1952). In practice, this severe view is somewhat modified if the employer terminates the contractor's employment under the terms of the contract (see section 12.1.7).

The contract sum may be adjusted only in accordance with the provisions of the contract (see Table 11.1). Then it becomes the contract sum which has been adjusted; the new amount is not the contract sum, which never changes. Errors or omissions of any kind in the computation of the contract sum are deemed to have been accepted by the employer and contractor (clause 4.2). The only exceptions to that are those instances specifically provided for (see section 3.1.3). They are inconsistencies in or between documents; errors or omissions in description or quantity; and departures from the Standard Method of Measurement. The contractor may make all kinds of errors in pricing the contract documents. It may underprice or overprice items, overlook items, or simply make a mistake in adding up totals. Once the price is accepted, however, it may not be altered. The quantity surveyor will have checked the contractor's calculations before the tender was accepted, but may not have noticed the error. It is always bad, from the point of view of both

Table 11.1
Contractual provisions regarding adjustment of the contract sum (including adjustments in the schedules)

Clause	Provision
2.3	Additions for fees and charges
2.6.2	Additional premium for early use by employer
2.9	Deductions for setting out errors not to be amended
2.14	Instructions – exceptions
2.30	Deductions for defects not to be made good
3.9	Deductions in respect of employer's costs after contractor's non-compliance with instructions
3.14	Additions to cover the cost of opening up or testing if the work is in accordance with the contract
4.2	General provision governing adjustment
4.13	Final adjustment of the contract sum
4.15	Taking account of fluctuations
4.16	Taking account of fluctuations in respect of named sub-contractors
4.17	Direct loss and/or expense
6.3.3	Premiums for insurance of liability of the employer
6.13.2	Deduction to cover employer's costs for remedial measures in respect of the Joint Fire code
Schedule 1 para A.5.1	Difference in premium as a result of terrorism cover
Schedule 1 para B.2.1.2	If the employer defaults and the contractor has to take out insurance
Schedule 1 para C.3.1.3	If the employer defaults and the contractor has to take out insurance
Schedule 2 para 10.1	Following termination of employment of a named sub-contractor

contractor and employer, if the contractor finds, after entering into the contract, that it has made an error which will result in loss of money. The contractor will naturally attempt to recoup its losses by taking every opportunity to submit claims. So particular care must be exercised, before a tender is recommended for acceptance, that the checks made by the architect or the quantity surveyor have been thorough and that the final figure does not appear suspiciously low in comparison with other tenders.

11.2 *Payment before practical completion*

11.2.1 Method and timing

The parties may make whatever arrangements they wish for interim payments. Where the contract is of relatively high value, it is customary to pay at monthly intervals, but if the value is low or the priced documents make it convenient, it may suit both parties to agree that payment will be made on the completion of certain defined stages. The parties may have the agreement set out in the specification, schedules of work or bills of quantities before entering into the contract, or they may agree the mode of payment before commencing work. If a particular system of payment is desired, it is best to have it set out in the contract documents at tender stage because:

- The method and regularity of payment will significantly influence the contractor's tender; and
- If no agreement to the contrary is concluded after the parties have entered into a contract, the provisions of clause 4.6 will apply.

The contractual provisions are straightforward, but certain points need careful study.

Payment is to be made after the architect issues a certificate. The certificates are to be issued on the dates set out in the contract particulars. If no dates are inserted, the default position is to be one month (i.e. one calendar month). The intervals are calculated from the date of possession. In the case of deferment, it is suggested that, to avoid absurdity, the intervals are to be calculated from the new or deferred date of possession. The architect's certificate must include the total of the amounts as specified in clauses 4.7.1 and 4.7.2 less the total of the amounts in clause 4.7.3 at a date not more than seven days before the date of the certificate less the amount of any advance payment due for reimbursement and less any previously certified amounts. The employer has 14 days from the date of issue of the certificate to make payment.

If a calendar month is some four weeks, the default position is that the contractor could wait six weeks from the date of possession before receiving its first payment for three weeks' work. The importance of prompt payment cannot be overemphasised. It is a means of assisting the contractor's cash flow, reducing its overdraft requirements and hence the interest it has to pay, and so increasing its chance of making a reasonable profit, thus making the resort to the submission of claims less likely.

11.2.2 Valuation

The quantity surveyor must carry out interim valuations whenever the architect feels that it is necessary to do so (clause 4.6.2). The clause says 'necessary for the purpose of ascertaining the amount to be stated as due'. To ascertain is to find out for certain, so this step will be taken whenever the architect feels unable to know for certain without the aid of the quantity surveyor. Because the process of valuation is specialised, there will be few instances, in practice, when the wise architect will not require the quantity surveyor's assistance. Indeed, the contractor may force the quantity surveyor to prepare a valuation by submitting its own valuation not later than seven days before the date of the next certificate. In making an interim valuation, the quantity surveyor must notify the contractor, by means of a detailed statement, of any items of disagreement with the contractor's valuation. The clause requires the quantity surveyor's statement to be in similar detail to the contractor's application.

The amount to be included on the architect's certificate is to be the total of amounts in clauses 4.7.1 and 4.7.2, less the total amounts in clause 4.7.3 less any amounts included on previous certificates: note, not any amounts previously paid, but any amounts previously certified – and less any advance payment due for reimbursement to the employer and stated in the contract particulars under the terms of clause 4.5. Therefore the architect need not be concerned, when certifying, about whether the employer has paid or paid in full. The amounts to be included are divided into two categories: those items on which retention money will be retained; and those whose full value must be included (i.e. no retention).

The items are discussed in section 11.2.3. In the event of the contractor's insolvency, it is essential that the architect has not overcertified: *Sutcliffe* v. *Thackrah* (1974). Certification is the architect's responsibility whether or not a valuation has been provided by the quantity surveyor. An architect who disagrees with the quantity surveyor's valuation has a duty to change it. The architect must ensure that the quantity surveyor puts no value on defective work. The quantity surveyor cannot be expected to know what the architect considers to be defective unless the architect gives proper notice; this should be done in writing every month before the valuation is carried out.

Clause 4.5 deals with advance payment. The contract particulars state that it does not apply if the employer is a local authority. It is entirely a matter for the employer whether to operate the advance payment provision. If the employer decides to make advance payment, the contract particulars must be completed accordingly. The amount of payment must be inserted, and the times for repayment and the amount must be stated. If a bond is required this must also be stated. The advance payment is to be made before the issue of the first certificate for payment. No doubt the employer will normally pay it to the contractor after the contract is executed. If the employer requires a bond, and it is difficult to see why a bond would not be required, the terms must be those agreed between the British Bankers' Association and the Joint Contracts Tribunal. IC and ICD helpfully reproduce the text of this bond and of the appropriate notice of demand. There will clearly be many instances where a contractor will welcome money 'up front' to assist in setting up the site and funding the start of work and receipt of materials. If the repayment schedule is carefully worked out, the repayments should be covered by the monthly certificates. Obviously, the employer will expect a price advantage when opting to effectively fund the contractor in this way. The bond makes provision for repayment of the money to the employer on demand.

11.2.3 Amounts included

It is logical to consider the amounts to be included in the architect's certificate separately in the two categories. Amounts on which the employer is allowed a retention are stated in clause 4.7.1 to be:

- The total value of work properly executed by the contractor including valuations agreed under clause 5.2.1 or which have been valued in accordance with the valuation rules in clauses 5.3 to 5.6 inclusive, but excluding any restoration or repair of loss or damage and removal and disposal of debris treated as a variation (paragraphs B.3.5 and C.4.5.2 of schedule 1). Where the contract particulars state that a priced activity schedule is attached, one arrives at the value of such work by applying the price for each activity to the percentage of work in that activity properly executed.

- The total value of materials which have been reasonably and not prematurely delivered to, or adjacent to, the Works for incorporation, provided that they are adequately protected against weather and damage. The certificate need not include any materials which the contractor has clearly delivered to site for the express purpose of obtaining payment. Clause 2.17, unfixed materials on site, and clause 3.6.2, sub-contracting, are intended to ensure that materials paid for in this way by the employer become the property of the employer in law (clause 4.7.1.2). Whether they are successful will very much depend on whether the materials are the sub-contractor's property in the first instance.
- The value of any off-site materials on the employer's list in accordance with clause 4.12. The inclusion of off-site materials in a certificate has the potential to raise serious problems. Clause 2.18 is intended to ensure that, once the employer has paid for them, the materials become the employer's property. The contractor is not permitted to remove the materials, nor to allow anyone else to do so, except for use on the Works. All the while, the contractor is to remain responsible to the employer for any loss or damage. There are two dangers. First, the supplier may have incorporated a retention-of-title clause in the contract of sale to the contractor. That means that, despite anything that might be written into this contract, the materials remain the property of the supplier until payment is received from the contractor. Compare these provisions with the provisions for sub-contractor's materials in clause 3.6.2 (see section 8.2.1). The second danger is that it is difficult to be sure that the materials inspected at the contractor's yard are not really intended for some other job. It has been known for a contractor to label and set aside, say, sink units for a particular contract until after inspection by the architect, then re-label them for the benefit of another architect and a different contract. This is sharp practice, but, if the contractor goes into liquidation, that is no consolation for the employer.

Clause 4.12 is in an attempt to safeguard the employer against the danger of paying for goods or materials to which the employer cannot subsequently prove ownership. The clause sets up a system for payment. If the employer wishes to pay the contractor for goods or materials stored off-site they must have been listed, the list must have been provided to the contractor (obviously at tender stage), and it must have been included with the bills of quantities, the specification or the schedules of work. An appropriate note must be made in the contract particulars. Once that has been done, the contractor is entitled to have the value of the listed items included in an interim certificate if certain conditions are satisfied. The conditions are:

- The contractor must have provided reasonable proof of ownership of the listed items. This is not an easy thing to do. Essentially the best that can be done is to be as sure as possible that there is nothing that suggests that the contractor does not own the items. For example, the contractor's sub-contract with the sub-contractor could be provided, together with a statement from the sub-contractor that the contractor has complied with any preconditions that need to be fulfilled before ownership is transferred. Whether the sub-contractor has secured ownership from its supplier is another question. The listed items are divided into two categories: uniquely identified items, such as specialist items, e.g. lift equipment or special boilers or purpose made furniture; and items which are not uniquely identified, such as bricks, tiles, or timber. In the latter case, the contractor must have provided a bond in terms set out in schedule 3. Where the item is uniquely identified, the matter of the bond is left to the employer's discretion. In either case, the contract particulars must state the amount for which the bond is required. It is difficult to envisage why the employer should not require a bond in all circumstances. In practice, however, payment for off-site materials is becoming something of a rarity.
- The listed items must be in accordance with the contract. Put simply, they must not be defective.
- Wherever the items are situated off-site, they must be set apart from other goods or materials, or they must be clearly marked with a reference and identity the employer and the name of the person who ordered them (usually this will be the contractor) and the destination as being the Works.
- The contractor must provide the employer with reasonable proof that the items are insured for full value under a policy which protects employer and contractor in respect of specified perils. The policy must be effective to cover the whole period from the commencement of the contractor's ownership to delivery to or adjacent to the Works.

Amounts on which the employer is allowed no retention are in accordance with the following:

- Clause 2.3, where the contractor has paid statutory fees or charges which were not provided for in the contract documents
- Clauses 3.14 and 3.15.3, where the contractor has carried out opening up and/or testing and the work is found to be in accordance with the contract; the contractor is entitled to be paid the cost of the work and the cost of making good
- Clause 4.15, contribution, levy and tax fluctuations
- Clause 4.16, named-person fluctuations
- Clause 4.17, where the contractor is entitled to payment of direct loss and/or expense due to disturbance of regular progress
- Clause 2.6, where the employer requires use or occupation and an additional premium is required by the insurers
- Clause 6.5.1, where the contractor has taken out and maintained special insurance which is the liability of the employer
- Paragraph A.4.4 of schedule 1, where the contractor has insured the works and an insurance claim has been accepted
- Paragraph B.2.1 of schedule 1, where the employer, not being a local authority, has failed to insure and the contractor can produce evidence that it has taken out the appropriate insurance itself
- Paragraph B.3.5 of schedule 1, where the contractor has restored loss or damage
- Paragraph C.3.5 of schedule 1, where the employer has failed to insure and the contractor can produce evidence that it has taken out the appropriate insurance itself
- Paragraph C.4.5 of schedule 1, where the contractor has restored loss or damage and the employer has insured.

The amounts payable will depend on the amounts ascertained at the time of the valuation. Deductions or withholdings are to be made in this category in accordance with the following:

- Clause 2.9, where the architect has instructed that errors in setting out should not be amended and an appropriate deduction should be made
- Clause 3.9, where the contractor has not complied with an instruction and the employer has had to engage others
- Clause 4.15, contribution, levy and tax fluctuations as appropriate
- Clause 4.16, named-person fluctuations as appropriate.

The failure to include the provision in clause 2.30 – that if the architect instructs that defects, shrinkages and other faults are not to be made good, an appropriate deduction is to be made from the contract sum – has been carried over from IFC 98.

The certificate must state to what the payment relates and the basis of calculation. Under the provisions of clause 4.8.2, and in order to comply with the Housing Grants, Construction and Regeneration Act 1966, the employer must give the contractor written notice, not later than five days after the date of issue of an interim certificate. The written notice must specify the amount of payment to be made, to what it relates and, once again, the basis of the calculation. It will be of little consequence if the employer forgets to give the notice provided the certificate is paid in full within the 14-day period set out in the contract.

11.3 Payment at practical completion

Clause 4.6.1 stipulates that regular interim certificates and payments will be made until practical completion. Although it is nowhere expressly stated, at that point they will cease, simply because there will be no further work to certify. It may be that a particularly difficult claim will not be settled until a month or two after practical completion. The architect should make certain that, in such a case, money is released to the contractor whenever it is decided that any part of the claim is valid. There is no contractual liability to do so, however, and the architect may wait until the whole claim has been quantified before certifying. Under the current wording of the contract, it is arguable that the architect may have to wait until the issue of the final certificate.

A contractor whose claim is ascertained a month after practical completion is in a difficult position and likely to be kept out of its money for many months. Under clause 4.2 of IFC 98, the contractor could argue that a certificate should be issued when money was due. Although the current clause 4.6.1 does not preclude such issue, it does not require it. If the money is substantial and clearly due, the architect has power to certify it, and it would be a mean-spirited architect indeed who refused to do so. It may be significant that JCT has chosen not to call the certificate at practical completion the penultimate certificate, as is the case with the lower-value contracts MW and MWD.

The contract provides for a special payment to be made at practical completion (clause 4.9). The architect must issue a certificate within 14 days of the date of practical completion. The employer must pay, as before, within 14 days of the date of the certificate. The amounts to be included fall into the same categories (see section 11.2.3) as for interim certificates, and there is the same requirement for notices. There is an important difference, however. The architect must include the percentage stated in the contract particulars of the value of amounts in the first category. The default percentage is 97½% assuming that the default retention of 5% has been kept until practical completion. This has the effect that the employer releases half the retention that has been withheld. The remaining retention is held by the employer until the final certificate.

11.4 Retention

Retention is dealt with specifically by clause 4.10. The object of the clause is to safeguard the contractor's interest in the retention, but there are other implications. So it is curious that the initial provision excludes local authorities from its operation. This must be because the draftsman, while following accepted practice in assuming that a local authority will not become insolvent, overlooked the clause's broader application. Alternatively, it may have been to prevent the employer becoming a trustee, with all that implies.

Except where a local authority is involved, the employer is stated to be a trustee with interest in the retention to be fiduciary (which amounts to much the same). The employer is a trustee for the contractor. That means that the contractor has the right to insist that the retention fund be kept in a separate bank account clearly designated as held in trust for the contractor: *Wates Construction (London) Ltd* v. *Franthom Property Ltd* (1991). This safeguards the money if the employer becomes insolvent. Some contracts now have a special provision requiring the separate account.

Although the employer holds the retention in trust for the contractor, the clause states that there is no obligation to invest – that is, no obligation to make the best use of the money on behalf of the contractor or to return interest to the contractor on it. This provision appears contrary to statute. However, the point has yet to be decided by the courts.

The contractor's interest in the retention is said to be subject only to the employer's right to take money from it from time to time to pay amounts which the contract provides for the employer to deduct from sums due or to become due to the contractor. So this clause, which is not applicable to local authorities, is the only express term allowing the employer to use retention monies for other than the contractor's purposes. No doubt, in the complete absence of this clause, a term would be implied to allow the employer to deduct from the retention to make good the contractor's defaults. In the case of a local authority, the clause is not absent, but it is stated to apply only 'where the employer is not a local authority'.

A list of the contract provisions allowing deduction from amounts due or to become due to the contractor is given in Table 11.2.

11.5 Final payment

The contract lays down a strict time sequence for the events leading up to, and the issue of, the final certificate (see Figure 11.1). The contractor has a duty to provide the architect, or the quantity surveyor if the architect so instructs, with all the documents which are reasonably required for the final adjustment of the contract sum (clause 4.13). Figure 11.2 is a suitable letter. The contractor may send them either before practical completion or the last section completion certificate, but no later than six months afterwards. The architect is entitled to request the kind of documents required or to request particular documents, provided that the requests are reasonable in the circumstances.

Armed with this information, the quantity surveyor must prepare a statement of all the final valuations under section 5. A copy of all the computations to arrive at the finally adjusted contract sum, together with the statement, must be sent to the contractor within three months of receipt. But if the contractor is unreasonably late in sending its documents, it cannot expect the architect to adhere to this timetable.

The contract does not state that the contractor must agree the finally adjusted contract sum before the final certificate is issued. In practice it is customary to try to obtain agreement, and the contractor is usually sent two copies of the computations, one for it to sign as agreed and return. This is, no doubt, why the contract allows 28 days from sending the computations or issuing a certificate of making good under clause 2.31, whichever is the later, for the architect to issue the final certificate (clause 4.14.1). The contract unaccountably allows 28 days for the employer or the contractor, as the case may be, to pay. Clause 4.14.2 stipulates that the employer must give written notice to the contractor specifying the amount of

Table 11.2
Contract provisions which entitle the employer to deduct from any sum due or to become due to the contractor

Clause	Provision
2.23	Liquidated damages
4.8.3	Notice of withholding from interim payment
4.10	Withholding from retention
4.14.3	Notice of withholding from final payment
6.4.3	Contractor's failure to insure

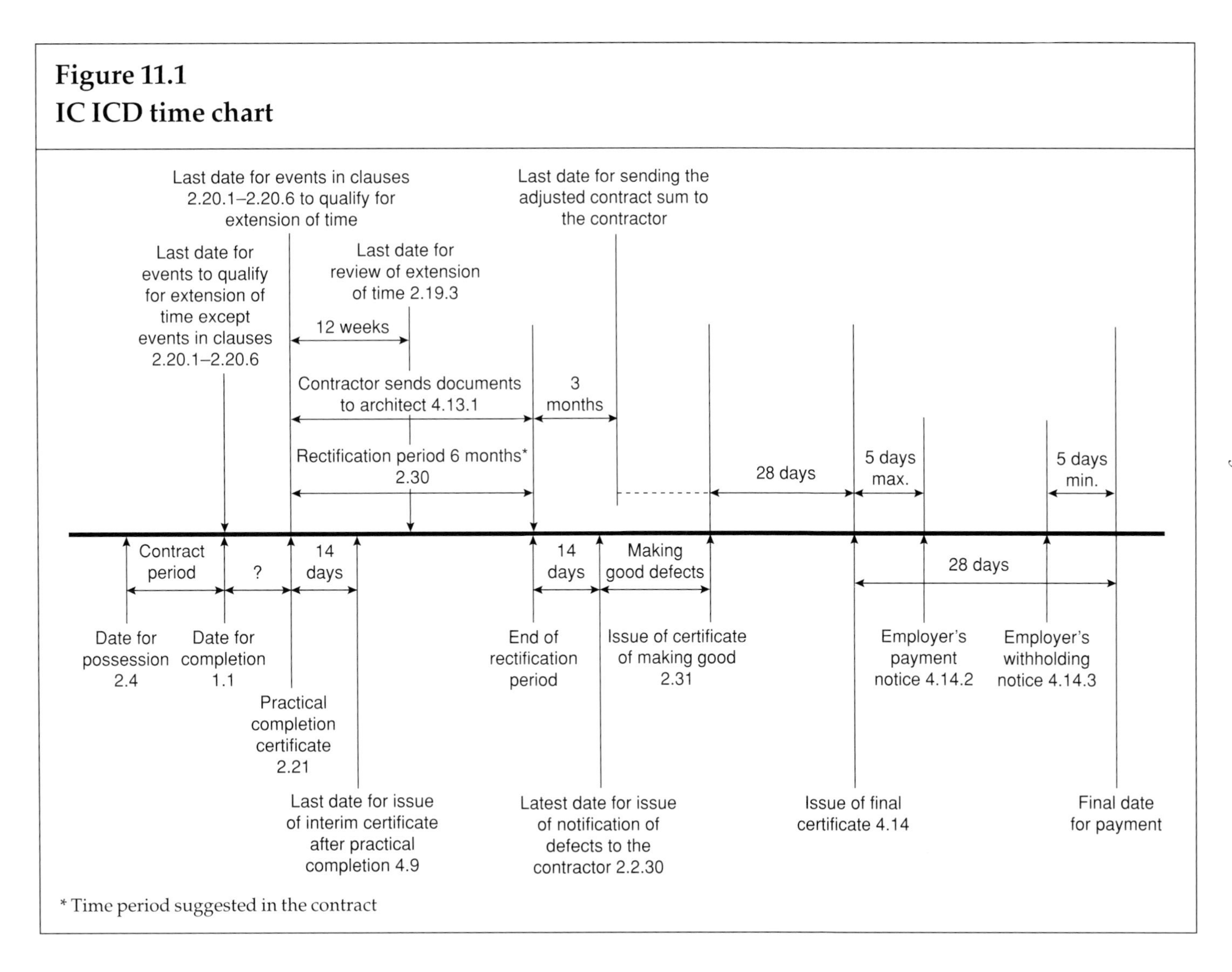
Figure 11.1
IC ICD time chart
Last date for events in clauses 2.20.1–2.20.6 to qualify for extension of time
Last date for sending the adjusted contract sum to the contractor
Last date for events to qualify for extension of time except events in clauses 2.20.1–2.20.6
Last date for review of extension of time 2.19.3
12 weeks
Contractor sends documents to architect 4.13.1
3 months
Rectification period 6 months* 2.30
28 days
5 days max.
5 days min.
Contract period
?
14 days
14 days
Making good defects
28 days
Date for possession 2.4
Date for completion 1.1
Practical completion certificate 2.21
Last date for issue of interim certificate after practical completion 4.9
End of rectification period
Issue of certificate of making good 2.31
Latest date for issue of notification of defects to the contractor 2.2.30
Employer's payment notice 4.14.2
Employer's withholding notice 4.14.3
Issue of final certificate 4.14
Final date for payment
* Time period suggested in the contract

Figure 11.2
Contractor to architect, enclosing all documentation for the preparation of final certificate

Dear

In accordance with clause 4.13.1 of the conditions of contract, we enclose full details of the final account for this contract together with all supporting documentation.

We should be pleased if you would proceed with the necessary calculations and verifications to enable the final certificate to be issued in accordance with the contract.

Yours faithfully

payment to be made. The notice must be given not later than five days after the date of issue of the final certificate. This notice provision serves a similar purpose to the notice provision for interim certificates.

The final adjustment to the contract sum must include:

- Any sums agreed for variations and all clause 5.2 valuations
- All amounts in accordance with clause 4.7.2 and deductions in accordance with clause 4.7.3
- Deduction of provisional sum amounts and work represented in the contract documents by approximate quantities
- Deductions under clause 2.30 (defects liability).

If the result is in favour of the contractor, the employer will be liable to pay; if the result is in favour of the employer, the contractor will be liable to pay. The latter is a somewhat unusual situation.

It has been held that the architect cannot refuse to issue the final certificate simply because the contractor has not provided the information. Neither that nor even the quantity surveyor's calculation of the final account is a prerequisite to the issue of the final certificate: *Penwith District Council* v. *VP Developments Ltd* (1999). If the architect fails to issue the final certificate in accordance with the contract timescale, the contractor should act immediately to secure its issue (Figure 11.3).

11.6 The effect of certificates

Clause 1.11 provides that no certificate, other than the final certificate, is conclusive evidence that any work, materials or goods to which it relates are in accordance with the contract. All certificates, except the final certificate, are included, whether financial or not. For example, the issue of the certificate of practical completion does not prevent the architect from requiring the contractor to make good work not in accordance with the contract. The issue of an interim certificate is not evidence that all the work included is in accordance with the contract. The architect is entitled to omit defective work from one certificate if it has been included in a previous certificate.

Under clause 1.10, the final certificate has a conclusive effect in four respects:

- Where, and to the extent that, approval of the quality of materials or of the standards of workmanship is expressly described in

Figure 11.3
Contractor to architect if final certificate not issued on time

SPECIAL DELIVERY

Dear

Clause 4.14.1 of the conditions of contract requires you to issue the final certificate within 28 days of the latest of the following events:

1. The sending to us of computations of the adjusted contract sum, which we received on [*insert date*].

2. Your certificate under clause 2.31 which was issued on [*insert date*].

Therefore the final certificate should have been issued, at latest, on [*insert date*]. Some [*inset number*] weeks have passed since that date and we have received no such certificate. You are in breach of contract, a breach for which, we are advised, the employer is liable. If the final certificate in the agreed amount is in our hands by [*insert date*], we will take no further action on such breach. Otherwise, we shall refer the dispute to adjudication, and we shall be seeking an order that the certificate be issued forthwith and interest paid on the amount which should have been certified.

Yours faithfully

Copy: Employer

a contract document or an architect's instruction to be for the architect's approval, the final certificate is conclusive that the quality and standards are to the architect's reasonable satisfaction. But there is an important restriction to the effect that the final certificate is not to be conclusive that the materials, goods or workmanship in question, nor any other, comply with any other term of the contract. In other words, the final certificate, in some circumstances, may be conclusive about the architect's satisfaction, but no more than that. It leaves open the question whether the employer could contend that the contractor would still be liable even though the architect was satisfied: *National Coal Board* v. *Neill* (1984). The important nature of this clause, in conjunction with clause 2.2, has been discussed fully in section 4.2.1, and the effect of the *Crown Estates* v. *Mowlem* case was taken into account by the draftsman of this clause:

- That the terms of the contract that require additions, deductions or adjustments to the contract sum have been correctly operated
- All due extensions of time have been given
- Reimbursement of loss and/or expense under clause 4.17 is in final settlement of all contractor's claims in respect of the relevant matters.

There are two exceptions:

- If there have been any accidental inclusions or exclusions of any items, or any arithmetical errors in any computation, they may be corrected
- If any matter is the subject of proceedings, including adjudication commenced before, or within 28 days after, the issue of the final certificate, the certificate is not conclusive regarding that matter. Thus either party has 28 days after the date of issue to refer to arbitration or to commence legal action through the courts. If the matter is not one about which the certificate is stated to be conclusive, the parties have the normal limitation periods of either six or twelve years in which to bring an action, depending on whether the contract is under hand or a deed respectively.

The conclusion is that, provided that nothing or very little has been left to be to the architect's satisfaction or approval, and provided that the quantity surveyor has done the adjustment of the contract sum correctly, the issue of the final certificate is very much in the employer's interest.

11.7 Withholding payment

If the employer fails to pay any amount due to the contractor by the final date for payment (i.e. 14 days from the date of issue of a certificate except for the final certificate), clause 4.8.5 in respect of the interim payment and payment on practical completion and clause 4.14.5 in respect of the final certificate require the employer to pay simple interest on the outstanding amount at the rate of 5% over the current base rate of the Bank of England at the time payment becomes overdue.

If the employer wishes to withhold or deduct an amount from a payment to the contractor, the employer must give a written notice to the contractor not later than five days before the final date for payment. The notice must state the grounds for withholding payment and the amount of money withheld in respect of each ground. The information must be detailed enough to enable the contractor to understand the reason why it is not receiving the amount withheld. In *Rupert Morgan Building Services (LLC) Ltd* v. *David Jervis and Harriett Jervis* (2004) the Court of Appeal gave a very strong hint that the architect may have a duty to advise the client about issuing such a notice. Even without knowing the court's views, there can be little doubt that the architect, or indeed any other person acting as contract administrator, would have such a duty.

The contractor may seek immediate adjudication under the provisions of article 7 (see Chapter 14).

11.8 Variations

Unless the contract is extremely simple, the valuation of variations is best left in the hands of the quantity surveyor. This attitude is adopted in IC and ICD, and clause 5.2 provides that the contractor and the employer may agree on the amount to be added to, or deducted from, the contract sum in respect of any variation. This simply states the general law position, because two parties to a contract can vary its terms in any way they both agree. Note that the contract expressly reserves the right of agreement to the employer, not to the architect.

If they do not agree on a price, the valuation is to take place in accordance with the valuation rules set out in clauses 5.3 to 5.6. In ICD, valuation of the contractor's designed portion is carried out in accordance with clause 5.7.

Usually, the valuation will be carried out by the quantity surveyor. The quantity surveyor's principal tool in carrying out valuations is what this contract refers to as the 'Priced Document' (see the fourth recital of IC and the fifth recital of ICD). The wide range of possible documents has been discussed in section 3.1.1. The priced documents may be any one of the following: the priced specification, or the priced work schedules, or the priced bills of quantities, or the contract sum analysis, or the schedule of rates.

Omissions are to be valued in accordance with the relevant prices in the priced document. That is crystal clear, and should cause no problems (clause 5.3.2). Additional work may be valued in one of five ways:

- Work of similar character carried out under similar conditions and which does not significantly change the quantity of that which is set out in the priced document: the valuation must be consistent with the values in the priced document. In this clause it is probable that 'similar' can be given its ordinary meaning of 'almost identical', because the rates and prices are to determine the valuation. Thus, if there is no change in the conditions under which the work is to be carried out and the quantity is not significantly changed, the prices in the document are to be used for the valuation. In considering 'similar conditions', the quantity surveyor is confined to the conditions as identified in the contract documents. He or she is not entitled to take into account the background against which the contract was made: *Wates Construction (South) Ltd* v. *Bredero Fleet Ltd* (1993).
- Where the conditions and quantity are changed, the valuation will begin to depart from the values in the priced document, which must remain the basis for valuation. There will clearly come a point at which the relationship with the priced document will be very tenuous indeed and, in effect, a fair valuation will result.
- Where work is not of similar character, fair rates and prices must be used for the valuation.
- Where an approximate quantity is included, and it is a reasonably accurate forecast of the quantity actually required, the valuation must be in accordance with values in the priced document.
- If the approximate quantity is not a reasonably accurate forecast, the values are to be the basis of the valuation, which must include a fair allowance for any difference in quantity.

The procedure is set out in a flowchart in Figure 11.4. In practice, the quantity surveyor will look at the variation and see whether or not it is of similar character to work included in the priced document. If it is, those prices will be used as a basis for the valuation. If not, a fair

Figure 11.4
Valuation of variations (clause 5)

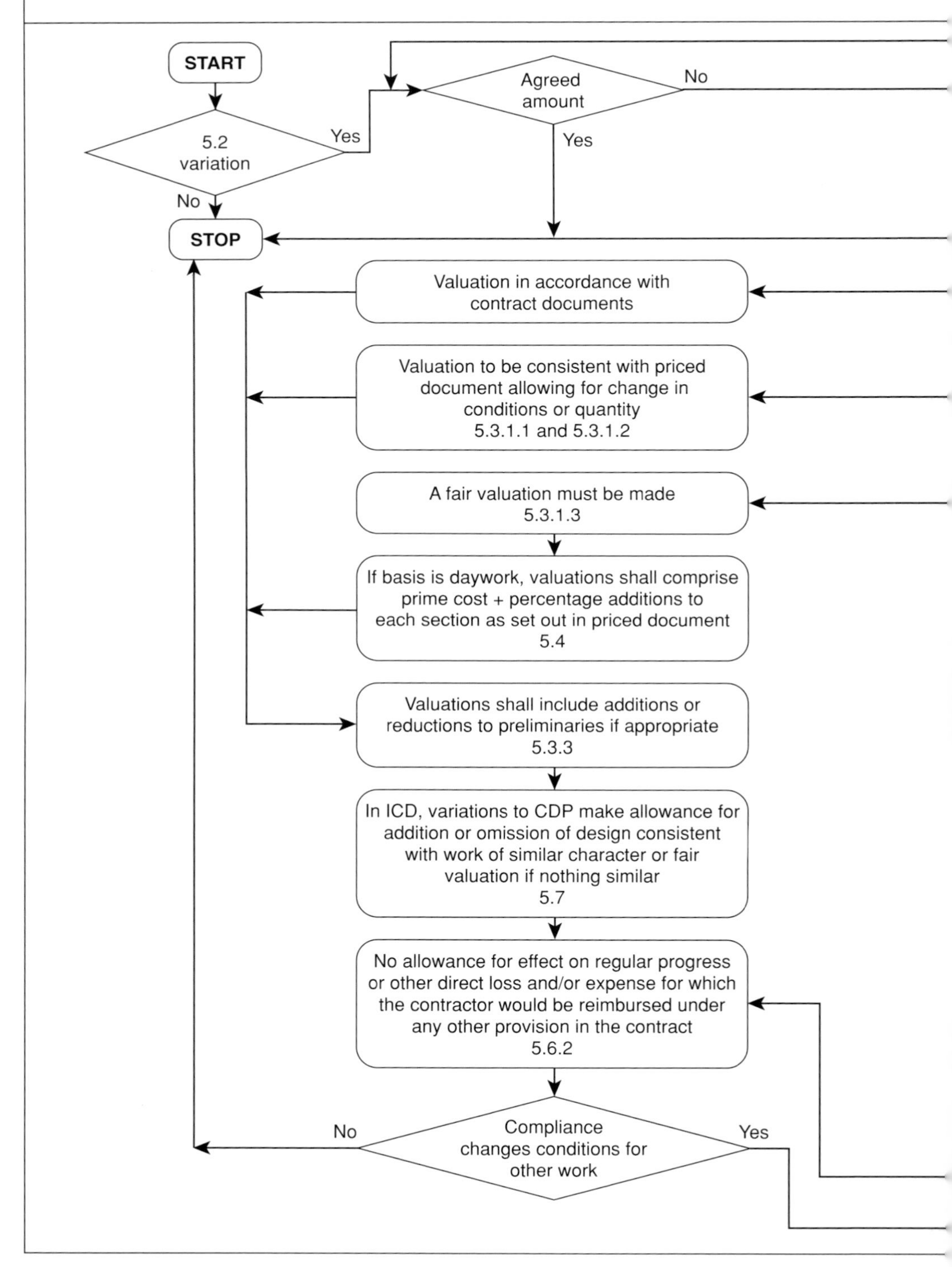

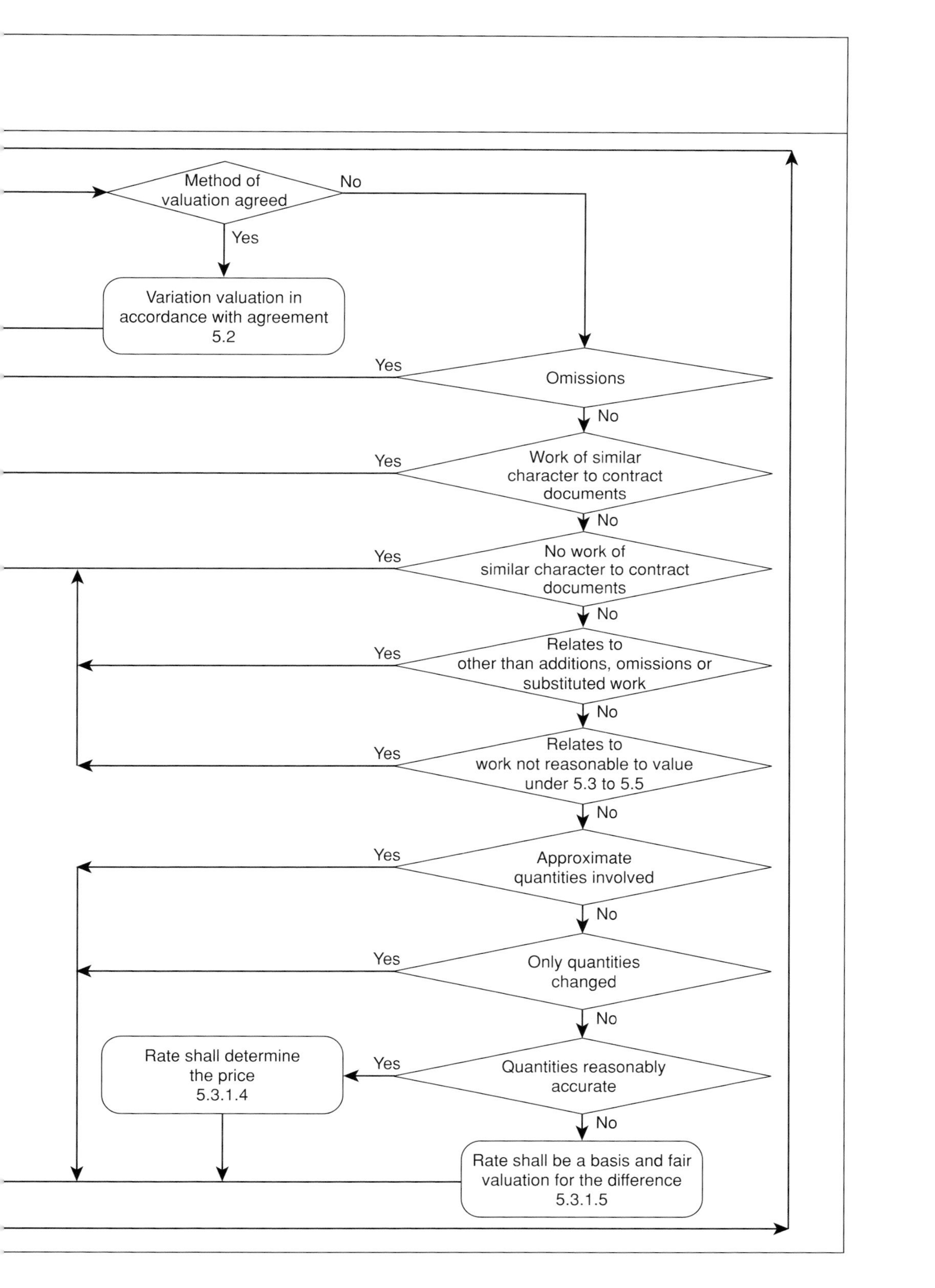

Method of valuation agreed
No
Yes
Variation valuation in accordance with agreement 5.2
Yes
Omissions
No
Yes
Work of similar character to contract documents
No
Yes
No work of similar character to contract documents
No
Yes
Relates to other than additions, omissions or substituted work
No
Yes
Relates to work not reasonable to value under 5.3 to 5.5
No
Yes
Approximate quantities involved
No
Yes
Only quantities changed
No
Rate shall determine the price 5.3.1.4
Yes
Quantities reasonably accurate
No
Rate shall be a basis and fair valuation for the difference 5.3.1.5

valuation will be made. The wording of the clauses probably gives the quantity surveyor considerable freedom.

If it is decided that the proper basis of any fair valuation should be daywork, clause 5.4 stipulates that the valuation must comprise the prime cost of the work together with percentage additions on the prime costs at the rates set out by the contractor in the priced documents.

Definitions of prime costs are many and varied. The contract sets out those that are acceptable:

- Generally prime cost is to be calculated in accordance with the Definition of Prime Cost of Daywork carried out under a Building Contract, issued by The Royal Institution of Chartered Surveyors and the Construction Confederation.
- If the work is within the province of any specialist trade and there is a published agreement between the RICS and the appropriate employers' body, prime cost is to be calculated in accordance with the definition in such an agreement. A footnote to clause 5.4.2 states that this subparagraph refers to three definitions: those agreed between the RICS and the Electrical Contractors' Association; those agreed between the RICS and the Electrical Contractors' Association of Scotland; and those agreed between the RICS and the Heating and Ventilating Contractors' Association. As the footnotes are not part of the contract, there is nothing to prevent other definitions of prime cost from being used, provided that they fall within the meaning of the subparagraph. The definitions to be used are those which were current at the date of tender.

There is an express requirement for the contractor to submit vouchers for verification to the architect or an authorised representative no later than the end of the week following the week in which the work has been executed. Contractors generally prefer the valuation to be done by means of daywork, so there should be no difficulty.

When valuing on a daywork basis, the quantity surveyor has no power to substitute his or her own opinion of the time and resources which would have been reasonable for those on the daywork vouchers or sheets: *Clusky (trading as Damian Construction)* v. *Chamberlain* (1995). Inscribing the words 'for record purposes only' does not prevent the sheets being used for the calculation of payment: *Inserco* v. *Honeywell* (1996). Where the sheets are properly submitted by the contractor, but the architect fails to sign them, they will stand as evidence of the work done: *J.D.M. Accord Ltd* v. *Secretary of State for the Environment, Food and Rural Affairs* (2004).

Additions or reductions to appropriate preliminary items must be included in the valuations except where compliance with an architect's instruction for the expenditure of a provisional sum for defined work in accordance with the Standard Method of measurement is involved (clause 5.3.3).

If the conditions under which any other work is carried out are substantially changed by reason of the contractor's carrying out of work in accordance with the architect's instructions, the other work must be treated as if it too had been varied. Two points should be noted:

- If an approximate quantity is included in the contract documents, this rule will apply to the extent that the actual quantity differs from the approximate quantity
- If bills of quantity are used, this rule will apply to the expenditure of a provisional sum for defined work to the extent that the instruction differs from the bill description (clause 5.5).

For example, the architect's instruction to divide a large warehouse with brick walls may make it more difficult to lay floor screeds in some areas. In such a case the quantity surveyor must value not only the cost of building the brick walls but also the cost of laying the floor screeds under different conditions. The change must be substantial, i.e. it must not be trivial. It is for the quantity surveyor to decide where to draw the line.

The quantity surveyor is not to make any allowance in the valuation for the effect of the architect's instruction on the regular progress of the Works or for any other direct loss and/or expense for which the contractor would be reimbursed by any other provision of the contract (clause 5.6). This is clearly intended to avoid confusion between this clause and clause 4.17 and to prevent the contractor from claiming, and being paid, twice for the same matter. It is still open, of course, for the contractor to satisfy the quantity surveyor that the effects on progress and loss and/or expense are not covered by other contract provisions. If the contractor can do this, it is entitled to be paid under this clause.

11.9 *Valuation of contractor's designed portion*

This is treated separately under clause 5.7 of ICD.

Clauses dealing with significant changes in condition and quantities (5.3.3.2), work not of similar character (5.3.3.3), daywork (5.4) and changes in condition of other work (5.5) are to apply to CDP work to the extent that they are relevant. Presumably it is for the architect or the quantity surveyor to decide that in the first instance, the contractor having power to seek adjudication or arbitration if it disagrees.

It is expressly stated that allowance must be made for addition or omission of relevant design work. This is considerably assisted if a rate for design is included in the CDP analysis. That might be on an hourly rate basis.

With crystal clarity, omission of CDP analysis work is to be valued in accordance with the relevant prices therein.

Added or changed work follows similar rules to those set out in clause 5.3. The values are to be consistent with similar character work in the CDP analysis with appropriate allowance for change in the conditions under which the work is carried out or the quantity of the work. If there is no similar character work, there must be a fair valuation, presumably by the quantity surveyor.

11.10 Fluctuations

There are two clauses which deal with fluctuation: clauses 4.15 and 4.16. The first deals with fluctuations allowable to the contractor's work, the second with fluctuations in respect of named persons. The fluctuations referred to in clause 4.15 are contained in schedule 4.

Schedule 4, dealing with contribution, levy, and tax fluctuations, provides for the bare minimum of fluctuations to cover changes in statutory payments such as National Insurance contributions.

The amounts included in the contract sum in respect of named persons are to be adjusted by the net amounts which the named person is due to receive or allow in accordance with the applicable sub-contract (ICSub/C) fluctuation option. There is a proviso. It applies if the period for completion of the sub-contract works has been extended by reason of an act, omission or default of the contractor. In that case, any sum which would have been excluded, were it not for the extension of time, is to be excluded from the net amount mentioned above. This is because the ICSub/C clause 'freezes' the application of the fluctuations clause if the sub-contractor fails to complete on time. Therefore, if it is the contractor's fault that the sub-contractor failed to complete, it is still liable to pay the fluctuations to the sub-contractor, but the employer need not pay those amounts to the contractor. The sub-contract provisions are complex. For example, if the ICSub/NAM/C provisions for extension of time are amended in any way, or if the contractor does not respond to claims for extension of time within the prescribed period, the right to freeze fluctuations is lost. This is identical to the provisions to be found in the Standard Building Contract SBC. It is thought that, in such circumstances, the proviso to clause 4.16 would be robbed of all effect, whether the amendment was carried out by the employer or by the contractor.

11.11 Summary

Contract sum

- The contract sum is the amount for which the contractor agrees to carry out the whole of the Works
- IC and ICD are 'lump-sum' contracts
- A contractor who abandons the work is entitled to nothing
- The contract sum can be adjusted only in accordance with the provisions of the contract
- Once the contractor's price has been accepted, it is fixed.

Payment before practical completion

- The parties may make their own arrangements
- If there is no agreement to the contrary, clause 4.6.1 applies
- The default interval is one month between certificates
- The employer has 14 days in which to pay
- Employer must give five days' notice of amount to be paid
- A valuation may be carried out by the quantity surveyor not more than seven days before the certificate
- The employer is not allowed a retention on all amounts.

Payment at practical completion

- The certificate must be issued within 14 days of the date of practical completion
- The employer must give five days' notice of the amount to be paid
- Half the retention must be released
- Unless the employer is a local authority, it acts as trustee for the retention fund
- The contractor has the right to require the retention fund to be kept in a separate bank account
- The employer, if not a local authority, has the right to take out money to pay amounts that the contract provides for the employer to deduct.

Final payment

- The contractor must send all documents for computing the final adjusted contract sum
- They must be received no later than six months after practical completion
- A copy of the quantity surveyor's final computations must be sent to the contractor within three months of receipt of contractor's information
- The final certificate must be issued within 28 days from either the date the computations are sent, or the date of the certificate of making good, whichever is the later
- The employer must give notice not later than five days after issue of the final certificate.

Effect of certificates

- No certificate is conclusive evidence that work or materials are in accordance with the contract
- The final certificate is conclusive that, where it is expressly stated that anything is to be to the satisfaction of the architect, the architect is satisfied; that terms of the contract requiring additions, deductions or adjustments to the contract sum have been correctly operated; that extensions of time have been given; and that loss and/or expense has been reimbursed
- The final certificate is not conclusive if items have been accidentally included or omitted or there are arithmetical errors, or if either party commences proceedings before 28 days after the date of issue.

Withholding payment

- Interest is payable on amounts overdue
- If the employer wishes to set off, written notice must be given in due time.

Variations

- The value may be agreed between the employer and the contractor
- Otherwise, the quantity surveyor must value in accordance with clauses 5.3 to 5.6
- Valuation may be based on the priced document or be a fair valuation
- The basis for a fair valuation may be daywork

- Acceptable definitions of 'prime cost' are indicated in clause 5.4
- Adjustment to preliminary items must be included
- Work changed by a variation to other work must be treated as a variation
- There is no allowance for loss and/or expense unless the contractor cannot otherwise recover.

Fluctuations

- Schedule 4 covers bare minimum statutory changes
- Named person fluctuations are to be paid net
- If sub-contract works are extended through the contractor's fault, no fluctuations are paid for that period.

CHAPTER TWELVE
TERMINATION

12.1 *Termination by the employer*

12.1.1 General

Termination is one of those things which is best avoided. If it is impossible to avoid, it must be done properly, or the consequences will be unpleasant for the employer. The whole procedure is surrounded by difficulties and pitfalls for the unwary. Among them are the following.

If termination is properly carried out, the employer will be faced with a project to finish with the aid of another contractor. In theory, the employer can recover all the costs from the first contractor, but of course the employer cannot recover the time lost. Even the recovery of costs is likely to be uncertain unless the amount of retention is greater than the amount to be recovered, when a simple deduction can be made. If termination is not properly carried out, the contractor may be able to bring an action for damages for unlawful repudiation of the contract. However, it has been held that it is not repudiation where a party has honestly relied on a contract provision, even though mistaken: *Woodar Investment Development Ltd* v. *Wimpey Construction UK Ltd* (1980). Many of the grounds for termination may give rise to dispute.

The employer will look to the architect for advice on whether or not to terminate the contractor's employment. Indeed, it will probably be the architect who brings the matter to the attention of the employer. This will often be because the situation on site has deteriorated to such a stage that the architect is pessimistic about the chances of ever achieving completion. Ideally, termination should be set in motion before that stage is reached, but in practice it is difficult to decide just when there is no hope of recovering the situation. These provisions are without prejudice to the employer's other rights and remedies (clause 8.3.1).

The procedure for termination is set out in a flowchart in Figure 12.1. Termination is dealt with in section 8, which covers termination

Figure 12.1
Termination by employer (clauses 8.4–8.6, 8.11)

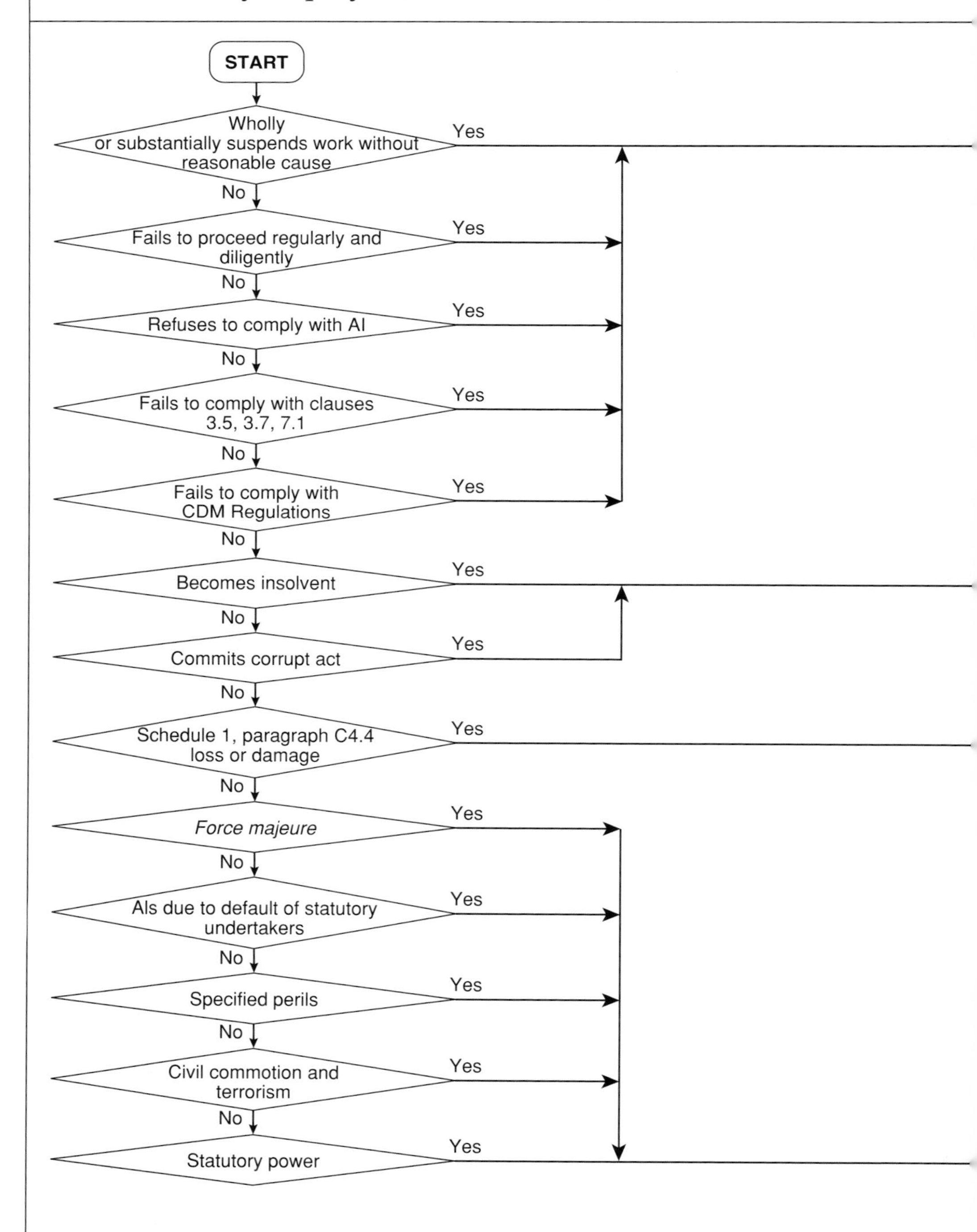

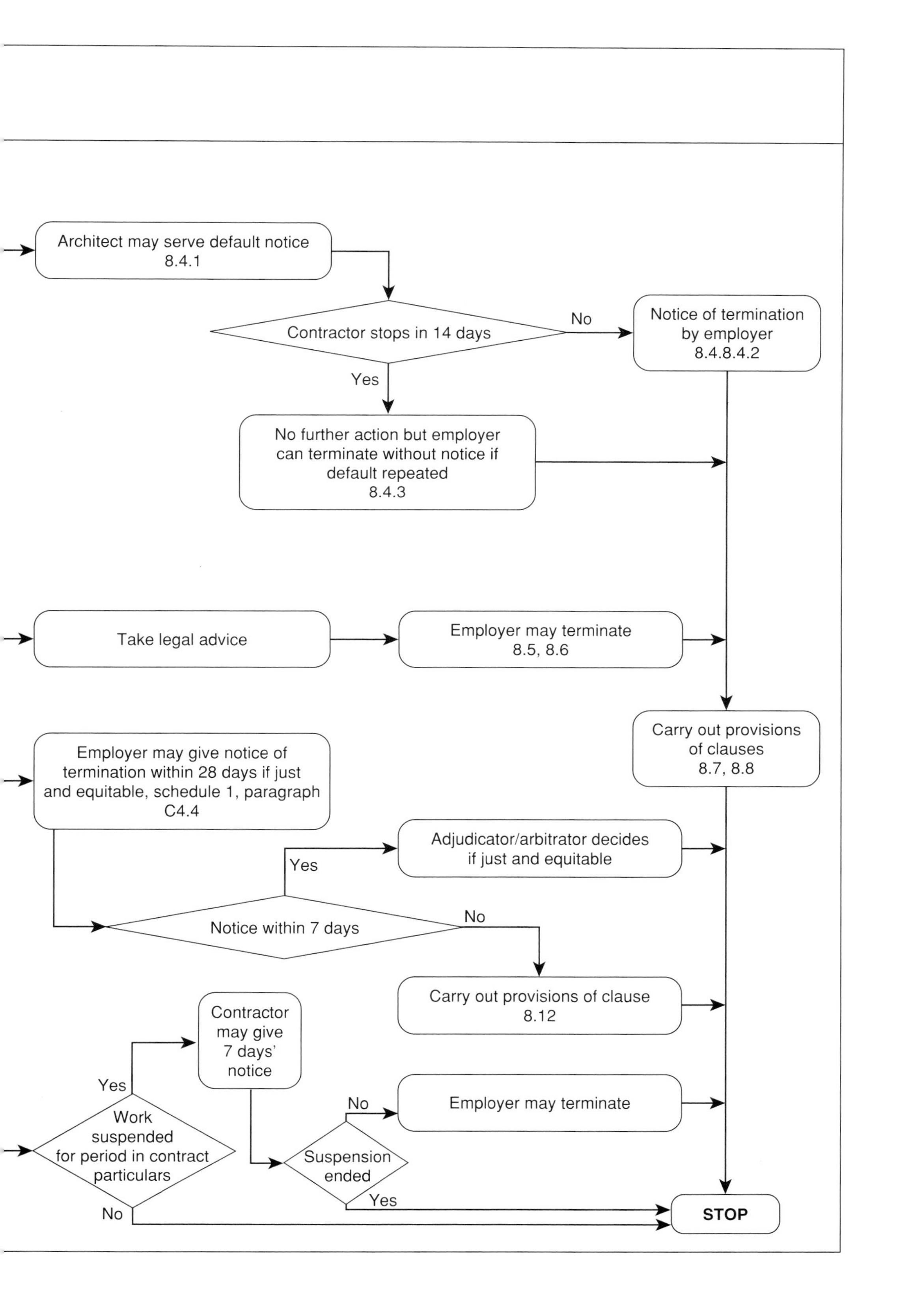

Architect may serve default notice 8.4.1
Contractor stops in 14 days
No
Notice of termination by employer 8.4.8.4.2
Yes
No further action but employer can terminate without notice if default repeated 8.4.3
Take legal advice
Employer may terminate 8.5, 8.6
Carry out provisions of clauses 8.7, 8.8
Employer may give notice of termination within 28 days if just and equitable, schedule 1, paragraph C4.4
Yes
Adjudicator/arbitrator decides if just and equitable
Notice within 7 days
No
Carry out provisions of clause 8.12
Contractor may give 7 days' notice
Yes
Work suspended for period in contract particulars
No
Employer may terminate
Suspension ended
Yes
No
STOP

by the employer, by the contractor, or by either party. A further and very specific provision for termination is included in paragraph C.4.4 of schedule 1. Notices must be given in accordance with clause 8.2.3, by actual, special or recorded delivery. Notice is deemed to have been received on the second business day after the date of posting. This provision may be useful, but it can be supplanted by evidence that the delivery was effected on a different day, and care should be taken. For example, special delivery can be guaranteed by the next business day, whereas recorded delivery may, on occasion, take several days to arrive. Clause 8 is a great improvement and simplification of the enormously complex clause 7 in IFC 98. The grounds for termination are set out in the contract in clauses 6.10.2.2, 8.4, 8.5, 8.6, 8.9, 8.10, 8.11 and paragraph C.4.4 of schedule 1. The consequences of termination are set out in clauses 6.10.3, 8.7, 8.8, 8.12 and paragraph C.4.4.2 of schedule 1.

12.1.2 Grounds (clause 8.4.1): contractor's defaults

There are five separate grounds for termination in clause 7.2.1. They are that the contractor:

- Wholly or substantially suspends the carrying out of the Works, before completion, without reasonable cause; or
- Fails to proceed regularly and diligently with the Works; or
- Refuses or neglects to comply with a written notice from the architect requiring it to remove defective work or improper materials or goods, and thereby the Works are materially affected; or
- Fails to comply with clauses 3.5 (sub-contracting), 3.7 (named persons), or 7.1 (assignment); or
- Fails to comply according to the contract with the CDM Regulations.

If the employer decides to terminate the contractor's employment, the procedure must be followed precisely. The contractor must be served with a notice of default (letter in Figure 12.2), which must clearly specify the default. It is very important that the architect says precisely what is wrong: *Wiltshier Construction (South) Ltd* v. *Parkers Developments Ltd* (1997). The architect must send the letter. It must be sent by special delivery, recorded delivery or by hand. It is wise to obtain a receipt for actual delivery. If the contractor continues the default for 14 days after receipt of the notice, the employer may terminate the contractor's employment by a further notice by special

Figure 12.2
Architect to contractor, giving notice of default

SPECIAL DELIVERY

Dear

I hereby give you notice under clause 8.4.1 of the conditions of contract that you are in default in the following respect: [*insert details of the default, with dates if appropriate*].

If you continue the default for 14 days from receipt of this notice, or if you at any time thereafter repeat such default (*whether previously repeated or not*), the employer may terminate your employment under the contract without further notice.

Yours faithfully

Copy: Employer
Quantity surveyor

delivery or recorded or actual delivery. It is usual for the architect to draft the letter for the employer's signature (Figure 12.3). The employer has 10 days to serve the termination notice. The period commences from the expiry of the 14 days.

A point sometimes arises concerning the date on which the first notice was received and thus from which the 14 days begin to run. The contract allows an assumption to be made about the date on which the notice would arrive in the ordinary course of the post, but if the assumption is wrong and the contractor can prove that the notice of termination was premature, the consequences may be serious. The wise course is for the architect to arrange for Royal Mail to confirm the delivery date.

There are two important provisos. First, if the contractor ceases its default within the 14 days, the employer can take no immediate action. If the employer has the right, but does not give notice of termination within the 10-day period and if the contractor repeats the same default at any time thereafter, the employer may terminate immediately or within a reasonable time without the necessity for a further 14 days' notice. This is a very powerful remedy in the hands of the employer.

Second, the notice of termination must not be given unreasonably or vexatiously. There must be no malice or intention to annoy: *John Jarvis* v. *Rockdale Housing Association Ltd* (1986). This is particularly applicable to the case where the contractor has stopped a default for some weeks or months but commits the same default again. Special care must be taken that a charge of unreasonableness cannot successfully be levelled at the employer.

Despite the provisions of the contract which say that the employer may issue a notice of termination without another notice, it would be prudent for the architect to send a warning letter (Figure 12.4), being careful to state that it is not a notice of default, because a further notice of default would be invalid: *Robin Ellis Ltd* v. *Vinexsa International Ltd* (2003). But a simple warning letter lets the contractor know that the employer intends to exercise the right to terminate. Usually, that will be sufficient to stop the default immediately. If it does not, the employer would certainly not be acting unreasonably in then terminating the contractor's employment. Great care must be taken that the latest default is the same as the original default. There is certainly scope for a contractor to challenge a termination on those grounds (Figure 12.5). If there is any doubt, the termination procedure should be set in train again.

Before the architect advises the employer to give notice, the five grounds must be carefully considered.

Figure 12.3
Employer to contractor, terminating employment

SPECIAL DELIVERY

Dear

I refer to the notice of [*insert date of original notice*] specifying the defaults.

In accordance with clause 8.4.2 of the conditions of contract, take this as notice that I hereby terminate your employment under this contract without prejudice to any other rights or remedies which I may possess.

The rights and duties of the parties are governed by clauses 8.7 and 8.8. The architect will write to you in due course with instructions regarding the temporary buildings, plant, tools, equipment, goods and materials on site. Subject only to your compliance with the architect's instructions you must give up possession of the site forthwith.

Yours faithfully

Copy: Architect
Quantity surveyor

Figure 12.4
Architect to contractor, giving warning of repeated default

Dear

This is not a notice of default under clause 8.4.1 of the conditions of contract.

A notice of default dated [*insert date*] was sent to you in respect of [*insert details of the default*].

It has come to my attention that the above default is being repeated. Under clause 8.4.3 the employer has the right to terminate your employment without further warning. I will visit the site tomorrow, and if the default is indeed being repeated, the employer intends to exercise the right to terminate without notice.

Yours faithfully

Copy: Employer
Quantity surveyor

Figure 12.5
Contractor to employer if employer terminates on the basis of repeated default

SPECIAL DELIVERY

Dear

We have today received your purported notice of termination dated [*insert date*] citing clause 8.4.3 and our alleged previous default which you specified in your default notice dated [*insert date*]. In our view this termination is invalid and you are not entitled to so terminate our employment under the contract.

We deny that we are in default as suggested or at all. Without prejudice to that contention, our default, if demonstrated, is certainly not a repetition of any previously notified default such as you specify.

Your attempted termination therefore amounts to a repudiation of the contract. As reasonable people, we are proceeding with the Works, and if we receive your formal withdrawal of notice of termination within three working days of the date of this letter, we are prepared to continue working normally. If we do not receive your withdrawal as aforesaid, we will pursue appropriate remedies, which will include substantial damages.

Yours faithfully

Copy: Architect

Wholly or substantially suspends the carrying out of the Works etc. (clause 8.4.1.1)

The contractor need not have completely ceased work. If a contractor, halfway through a £5 million contract, has only one or two men on site doing token work, probably it would not be regarded as having wholly suspended the Works, but it would certainly have substantially done so. This ground appears to be intended to cover the situation where the contractor has, in effect, abandoned the work or most of it. Note that the suspension must be without reasonable cause. Before advising the employer to send the initial notice of default, it may be prudent for the architect to ask the contractor why it had stopped. The course of action would depend on the reply. For example, the contractor might have suspended work for some reason that ranked for an extension of time. It might have suspended under clause 4.11.

Fails to proceed regularly and diligently with the Works (clause 8.4.1.2)

This ground implies more than simply failing to keep to a programme. The contractor's programme is not a contract document, although it may be a good indication of the contractor's intentions. 'Regularly and diligently' means that the contractor must work constantly, systematically and industriously. In deciding whether the contractor is working as required, regard must be had to such things as: the number of men on site, compared with the number of men required; the amount of plant and equipment in use; the work to be done; the time available for completion of the work; the actual progress being made; and factors outside the contractor's control (which may not all be clause 2.20 'Relevant Events') which hinder progress. It is clear that it is no easy matter to prove lack of regular and diligent progress. In *Greater London Council* v. *Cleveland Bridge & Engineering Co Ltd* (1986) a similar expression was considered, and the court was of the view that, provided the contractor carried out its work so as to meet any key dates and the completion date in the contract, it was not failing to exercise due diligence and expedition. It has been suggested that a contractor will be able to argue its way out of any termination on this ground if any progress at all is being made. However, the more recent Court of Appeal decision in *West Faulkner* v. *London Borough of Newham* (1994) quite firmly suggests otherwise (see section 5.1.2).

Refuses or neglects to comply etc. (clause 8.4.1.3)

Refusal to comply with the architect's instructions to remove defective work can be dealt with under clause 3.9, the employer engaging others to do the work. Note that JCT Amendment 3, issued in 1988, removed the word 'persistently' from before 'neglects'. The JCT Guidance Notes at the time suggested that the provision that notice must not be given unreasonably or vexatiously would prevent the use of trivial or one-off instances of neglect by the contractor being used improperly as a ground for termination, and that the deletion of 'persistently' would not therefore change the meaning of the clause. That is not a view shared by this author. One-off instances of neglect are covered in appropriate circumstances. In such circumstances there would be no question of the instance being used improperly, because it would be used in accordance with the contract provisions. This ground is probably intended to cover the situation where the contractor ignores instructions to such an extent that the work is in danger of grinding to a halt. It would also refer to the case where so much work is defective work that further work cannot be done without 'building in' the defective work and necessitating any future satisfactory work being taken down to make good the defective parts. The work on site would have to be in a very sorry state before termination should be attempted under this ground.

Fails to comply with clauses 3.5, 3.7 or 7.1 (clause 8.4.1.4)

Clause 3.5 refers to sub-contracting without consent. The object is to prevent the contractor from arranging vicarious performance of part of the contract by another. Because sub-contracting is traditional in the building industry, the contract makes provision for it provided that the architect consents. If the contractor does sub-let without consent, it is for the architect to decide whether the sub-contractor is suitable. If it is, there would seem to be no point, and little chance of success, in trying to terminate the contract. Even if the sub-contractor is clearly unsuitable, a less draconian method of dealing with the situation, probably by a letter (Figure 12.6), would probably be appropriate. Termination is best reserved for the occasions where the contractor sub-lets the whole or large portions of the work without the architect's approval.

Clause 3.7 refers to named persons. The provisions of this clause and schedule 2, to which it refers, are complex, but briefly this ground is aimed at giving the employer a remedy if the contractor

Figure 12.6
Architect to contractor if contractor sub-lets without consent

Dear

It has been brought to my attention that you have purported to sub-let [*insert the portion of the Works sub-let*] to [*insert name of purported sub-contractor*].

You have taken this course of action after I have refused consent/without asking my consent [*delete as appropriate*] to sub-letting. Since I have no intention of giving my consent, you are not permitted to use the above-mentioned purported sub-contractor on the Works.

I send this letter because I trust the incident is an oversight on your part, and I am reluctant to advise the employer to use rights under clause 8.4 if you will confirm to me, by return, that you will comply with this letter.

Yours faithfully

Copy: Employer
Quantity surveyor
Clerk of works

fails to sub-contract with a named person under any of the procedures or is otherwise in breach of its obligations under that clause.

Clause 7.1 refers to assignment. Neither party may assign the contract without the written consent of the other. If the contractor purports to do so, the employer may take steps to terminate. Attempting to assign the right to payment is clearly a serious matter.

Fails to comply with clause 3.18 (the CDM Regulations) (clause 8.4.1.5)

Essentially, this clause refers to a failure on the part of the contractor to comply with clause 3.18, which obliges it to comply with the duties of the principal contractor if the contractor takes that role and, if not, to comply with all the reasonable requirements of the principal contractor. The contractor must also provide – and ensure that each sub-contractor provides – information reasonably required for the health and safety file. Termination may seem a draconian remedy, but where health and safety is at stake and criminal penalties are severe, the employer must take decisive action.

12.1.3 Grounds (clause 8.5): insolvency of contractor

Insolvency is defined in clause 8.1. It includes:

- Entering into an arrangement, compromise or composition in satisfaction of debt
- A resolution for winding up
- A winding up or bankruptcy order made
- Appointment of administrator or administrative receiver
- Analogous arrangement, event or proceedings in another jurisdiction
- Individual arrangement or other event or proceedings as noted above affecting each partner of a partnership.

Even when a receiver or manager is appointed, it may be in the best interests of the employer to continue the contract. This can be decided only after a thorough discussion between all parties concerned. The employer would also be prudent to obtain legal advice regarding the implications in a particular case. The contractor's employment may be reinstated if agreement can be reached between the employer and the contractor. If the contractor is insolvent, the employer may at any time terminate the contractor's employment by written notice. Termination is effective on the date the notice is received by the contractor. Under clause 8.5.2, the contractor must immediately inform the employer if insolvency is likely.

12.1.4 Grounds (clause 8.6): corruption

The employer may terminate the contractor's employment if the contractor has given or received bribes in connection with this or any other contract with the employer, or if the contractor commits any other offence in relation to the contract or any other contract with the employer under the Prevention of Corruption Acts 1889 to 1916 or, if the employer is a local authority, under subsection (2) of section 117 of the Local Government Act 1972. A most onerous part of the provisions of this clause, so far as the contractor is concerned, is the fact that its employment may be terminated because of the corrupt actions of one of its employees or of some person acting on the contractor's behalf. It matters not that the contractor may have no knowledge of the affair.

In any case, corruption is a criminal offence for which there are strict penalties, and the employer is entitled at common law to rescind the contract and/or recover any secret commissions.

Legal advice is indicated, followed by a simple notice of termination, if that is the decision.

12.1.5 Grounds (clause 8.11): neutral causes

The employer (or the contractor) may terminate the contractor's employment if the carrying out of the whole or substantially the whole of the uncompleted Works is suspended for two months, or whatever period is entered in the contract particulars because of *force majeure*, architect's instructions regarding inconsistencies, variations or postponement issued as a result of negligence or default of a statutory undertaking, loss or damage to the Works caused by specified perils, civil commotion or terrorist activity or exercise by the UK government of statutory power directly affecting the Works.

For either party to operate this clause, the Works must be totally lacking in any significant progress for the entire period as a result of the same cause. One week of frenzied activity on the part of the contractor, in the middle of the period, may be sufficient to prejudice any attempt at termination, even if the site relapses into inactivity thereafter. It is probable, however, that the period must be viewed as a whole.

A seven-day period of notice is required at the end of the suspension of work. The notice must state that, unless the suspension is terminated within seven days of receipt of the notice, the employment of the contractor may be terminated. If the suspension does not end, the employer may, by further written notice, terminate the contractor's employment (Figure 12.7). There is an important proviso that the notice must not be given unreasonably or vexatiously.

All the causes of suspension are events beyond the control of the parties. They have all been discussed when considering extensions of time (see section 10.2.4). It is likely that both parties will be relieved to bring the contractor's employment to an end in such circumstances.

12.1.6 Grounds (paragraph C.4.4 of schedule 1 and clause 6.10.2.2): insurance risks and terrorism cover

This clause provides for the employer (or the contractor) to terminate the contractor's employment within 28 days of the occurrence of loss or damage to the Works or to any unfixed materials caused by any risks covered by the Joint Names Policy in paragraph C.2. The paragraph refers to work which is being done to existing structures by way of alteration or extension or both, but not to loss or damage to existing structures or contents. The contractor must give notice in writing to the architect and to the employer as soon as the damage is discovered. The notice must state the extent, nature and location of the damage. Although the 28 days begin to run from the occurrence and not from the notification, in practice if the damage is likely to be such as to form the basis for termination, it will be discovered and notified immediately it occurs.

A very important proviso in paragraph C.4.4 says, 'if it is just and equitable'. This proviso goes to the heart of the matter and points to the difference between this ground for termination and the ground in clause 8.11.1.3, which requires a long period of suspension. What is just and equitable depends on the particular circumstances. Despite the words at the beginning of paragraph C.4.1, 'If loss or damage', the sort of situation in which termination would clearly be just and equitable involves such catastrophic damage that it is uncertain not only when work could recommence but whether work could recommence at all. Take the case of a large factory, worth several million pounds. It may be that a small alteration and extension

Figure 12.7
Employer to contractor, giving 7 days' notice of termination after suspension

SPECIAL DELIVERY

Dear

The whole or substantially the whole of the uncompleted Works have been suspended since [*insert date*], a period of [*insert the period from the contract particulars*], by reason of: [*insert reason for suspension*].

In accordance with clause 8.11 of the conditions of contract, take this as notice that unless the suspension ceases within 7 days of the date of receipt of this notice I may, by further notice, terminate your employment under the contract. The termination will take effect on receipt of such further notice.

Yours faithfully

contract is let, worth £150,000. If, during the course of the work, the whole building is totally destroyed by fire, it will be just and equitable to terminate the contractor's employment. (In that case, the contract may also be considered to be frustrated at law.) Even much less than total destruction in such circumstances would give grounds for termination under this paragraph. The right of either party to seek adjudication or arbitration on the question of whether it is just and equitable is limited in two ways. The procedures must be invoked within seven days of receipt of a notice of termination, and the procedure is to decide whether termination will be just and equitable. It is, at best, doubtful whether this attempted restriction will be effective in the case of the right to adjudicate, which is backed by statute and exercisable at any time.

Clause 6.10 deals with the situation if terrorism cover ceases to become available. Clause 6.10.2.2 provides that, on receipt of a notice from the insurers, the employer may give written notice to the contractor stating that the contractor's employment will terminate on a stated date. The date must be after the insurer's notice, but before the date on which the terrorism cover is to cease.

12.1.7 Consequences (clauses 8.5.3, 8.7 and 8.8)

These clauses lay down the procedure to be followed after termination under clauses 8.4 to 8.6.

Clause 8.5.3 deals with the immediate effects if the contractor becomes insolvent. It is important to note that it is the contractor's insolvency which triggers these consequences and not any notice of termination. There are three consequences listed:

- The provisions regarding an account for the Works after completion and making good defects and if the employer does not wish to complete in clauses 8.7.3, 8.7.4 and 8.8 respectively are to apply as if notice of termination has been given, but other contract terms requiring further payment to the contractor or any release of retention will not apply. Therefore it seems that, even if an interim certificate has already been issued, the employer is not obliged to pay it. It is probable that this extends to the situation where the employer is actually already in default of payment.
- The contractor's obligation to carry out and complete the Works under clause 2.1 is said to be suspended, but there is no indication when such suspension might end. It is likely that the suspension is intended to take effect at insolvency, and notice of termination

from the employer would make the suspension a permanent cessation. The point of this clause is that, on insolvency, the contractor is not in breach of contract by failing to proceed with the Works.

- The employer is entitled to take reasonable measures to ensure that the site, the Works and materials on site are adequately protected, and to make sure that the materials are kept on site. This is because, on hearing of the contractor's insolvency, many unpaid suppliers will attempt to recover unfixed materials; whether they are entitled to do so will depend on many factors, and the employer must take specialist advice. The clause concludes by requiring the contractor to allow the employer to take these measures and not to hinder or delay them. In practice, if the contractor is insolvent, it is unlikely to take other than a passive role. A receiver or liquidator may potentially cause problems, and these provisions should go some way to preventing such problems from developing.

In the event that the contractor invokes dispute resolution procedures, the employer may deem it prudent to await the outcome, because if it is not possible to reinstate the original contractor, that contractor will be entitled to substantial damages (including loss of profit) if it wins.

Assuming that the contractor does not invoke dispute resolution procedures, or that the employer decides, nevertheless, to proceed, the consequences of the termination under clauses 8.4, 8.5 or 8.6 are as follows:

- The employer may employ another contractor to complete the Works, and the employer and the other contractor may take possession of the site. The completion of a partly finished building by another contractor is always an expensive procedure. In order to avoid a potential dispute, it is wise to have bills of quantities prepared for the completion work and to go out to tender in the normal way. This will prevent the original contractor from contending that the employer has not obtained a reasonable price.
- Although not expressly stated, the contractor must give up possession of the site of the Works. If the contractor does not do so within a reasonable time after receiving the notice of termination, it will become, in law, a trespasser, and the architect should send it a further notice to that effect (Figure 12.8). The contractor's liability for insurance ceases, and the employer must take out appropriate insurance cover without delay. This is best done at the time the notice of termination is sent, and the architect should

Figure 12.8
Architect to contractor if contractor refuses to give up possession of the site

SPECIAL DELIVERY

Dear

The employer terminated your employment under this contract in accordance with clause [*insert clause 8.4, 8.5 or 8.6*] on [*insert date*].

The consequences of termination are set out in clause 8.7. Sub-clause 8.7.1 provides that the employer and other contractors may enter upon and take possession of the site and the Works. On attempting to do so today, they were refused entry by some of your operatives. It is now [*insert number*] days since notice of termination was served on you and you have not given up possession. In law you are a trespasser, and if you have not given up possession of the site by [*insert date*] the employer intends to take whatever action is appropriate, including obtaining an injunction if necessary, to secure your removal. Obviously, you will be responsible for all the employer's legal costs.

Yours faithfully

Copy: Employer

remind the employer about this (Figure 12.9). There is no provision for the contractor to ensure that the Works are left in a safe condition, but the contractor (like anyone else) has a duty of care to those it can reasonably foresee could suffer injury. The contractor must not, therefore, leave walls or beams in a precarious condition.
- Clause 8.7.1 allows the employer to use any temporary buildings etc. for the benefit of a subsequent contractor. The architect should advise the employer whether use of the contractor's plant is desirable and may constitute a saving. If it is decided not to use it, the architect should instruct the original contractor to remove it from the Works.

Until the Works are complete, and defects during the rectification period are made good, the employer is not bound, and would not be wise, to make any further payment to the original contractor. Indeed, clause 8.7.2 expressly states that terms of the contract requiring payment or release of retention will cease to apply. When the contract is completed and the subsequent contractor paid in full, the employer must draw up a set of accounts, which must show:

- All the expenses and direct loss and/or damage caused to the employer by the termination. It will include the cost of completing the contract, including all professional fees consequent on the termination, the cost of engaging another contractor and, if appropriate, the cost of securing the site after insolvency
- The amount paid to the original contractor before termination
- The amount that would have been payable for the Works.

If the total of the first two amounts is greater or less than what would have been paid had the contract been completed in the normal way, the difference is a debt payable by the original contractor to the employer, or vice versa.

Invariably, the contractor owes a debt to the employer. The architect must carefully calculate, with the assistance of the quantity surveyor, the final amounts. It is essential that, on paper at least, the employer has been put in the same position as would have been the case had the contract not been terminated but had continued in an orderly way to its conclusion. Obtaining payment from a contractor which may be insolvent is another matter.

Under clause 8.8 the employer is given the option to elect not to complete the Works. This clause was introduced in response to *Tern Construction Group (in administrative receivership)* v. *RBS Garages* (1993), where the judge had to imply such a term in order to solve the

Figure 12.9
Architect to employer regarding insurance if contractor's employment terminated

Dear

Your notice of termination is being sent to the contractor today. The contractor no longer has any liability to insure the Works. You should consult your own broker without delay to obtain cover similar to that which the contractor was required to have under clause 6.4 and schedule 1 option A of the contract. Your insurance cover should be maintained at least until suitable arrangements have been made to complete the contract using another contractor.

Yours faithfully

problems in that case. The employer has six months from the date of termination to decide. If the employer opts not to complete, a written notification must be sent to the contractor. Within a reasonable time afterwards the contractor must be sent a statement (no doubt the architect or quantity surveyor will actually prepare it), which must set out:

- Total value of work properly executed at termination or at the date of insolvency, and any other amounts due to the contractor under the contract
- Expenses and direct loss and/or damage caused to the employer by the termination or otherwise.

The employer must take account of amounts previously paid and calculate the resultant balance due to the contractor or to the employer. The likelihood is that the employer will do rather worse under this system than if the employer opts to complete the work.

12.1.8 Consequences (clause 8.11 and paragraph C.4.4 of schedule 1)

The consequences of termination covered by these clauses are covered in section 12.2.7.

12.2 Termination by the contractor

12.2.1 General

If the contractor is successful in determining its employment under the contract, the results for the employer will be catastrophic. Among the consequences are these:

- The employer will be left with the project to complete with another contractor, completion bills of quantities must be prepared, and a great deal of additional expense will be incurred in the form of increased cost of completion and additional professional fees. The employer will be looking around to blame, and possibly take legal action against, someone, possibly one or all of the professional team.
- The completion date will be considerably exceeded.
- Under some of the grounds for termination, the contractor is entitled to receive loss of the profit it expected to make on the whole contract: *Wraight Ltd* v. *P.H. & T. (Holdings) Ltd* (1968).

The procedure for termination by the contractor is set out in a flowchart in Figure 12.10. The grounds for termination are set out in clauses 6.10.2.2, 8.9.8.10, 8.11 and paragraph C.4.4 of schedule 1. The consequences of termination are set out in clause 8.12. The provisions are without prejudice to the contractor's other rights and remedies (clause 8.3.1).

12.2.2 Grounds (clause 8.9): employer's faults

There are five separate grounds for termination in clause 8.9, as follows:

- The employer does not pay the contractor by the final date for payment the amount properly due under any certificate, including VAT
- The employer interferes with, or obstructs, the issue of any certificates
- The employer fails to comply with the assignment provisions
- The employer fails to comply according to the contract with the CDM Regulations
- The carrying out of the whole or substantially the whole of the works is suspended for a continuous period of two months or whatever period is stated in the contract particulars due to one of the following reasons:
  - architect's instructions under clauses 2.13, inconsistencies; or 3.11, variations; or 3.12, postponement, unless caused by contractor's default
  - any impediment or default of the employer, the architect, the quantity surveyor or any of the employer's persons.

If the contractor decides to terminate on any of the above grounds, the procedure must be followed precisely, otherwise the contractor may be simply attempting unlawful repudiation of the contract. The contractor must send a notice to the employer (not to the architect) by special or recorded or actual delivery (Figures 12.11 and 12.12), specifying one of the matters referred to in this clause. The contract differentiates between the first four grounds (under clause 8.9.1), which it calls 'specified defaults', and the fourth ground dealing with suspension (under clause 8.9.2), which it calls 'specified suspension events'. The difference appears to be based on grammatical rather than contractual necessities. They will all be referred to here as 'matters' for convenience. If the employer continues to make

Figure 12.10
Termination by contractor (clauses 8.9–8.11)

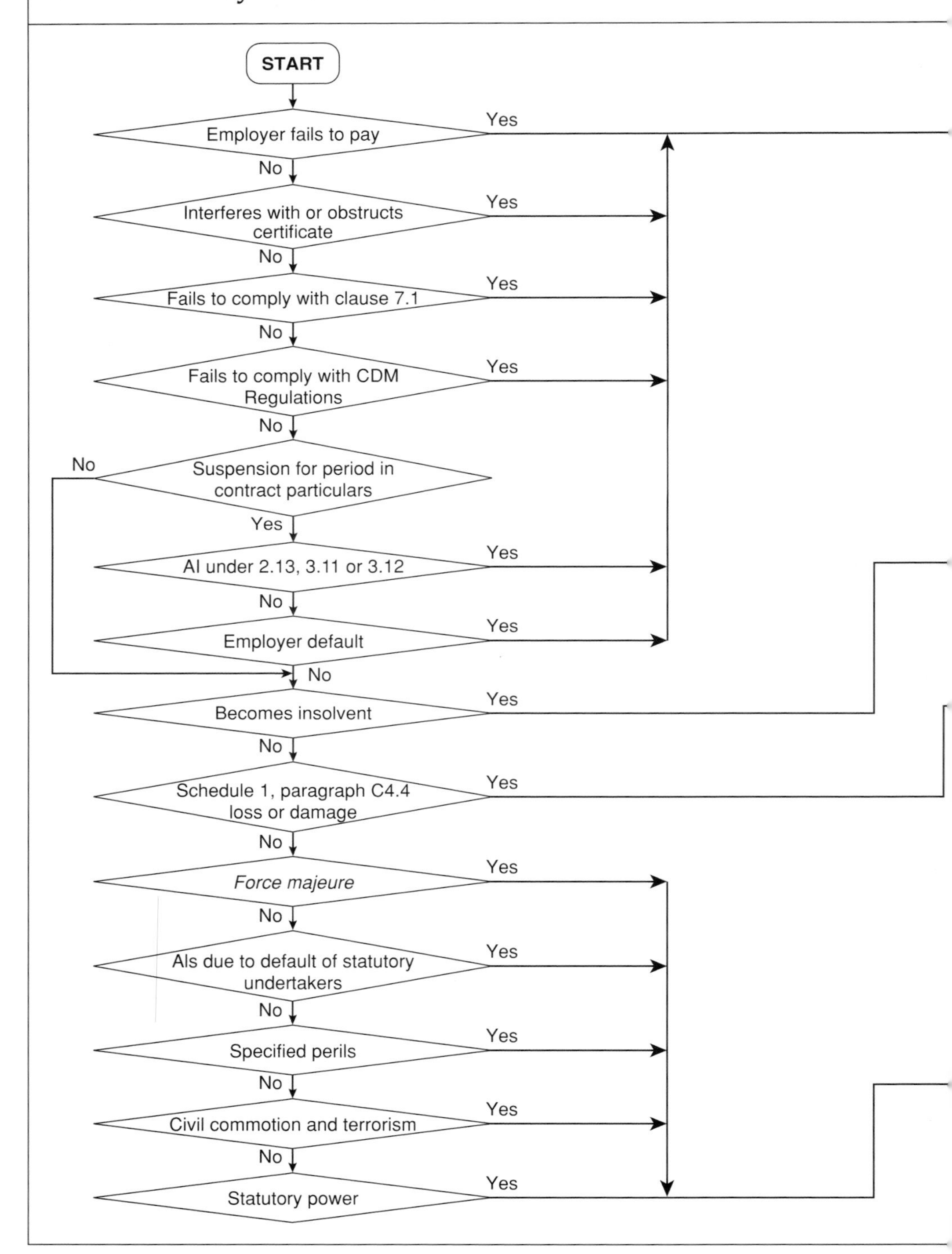

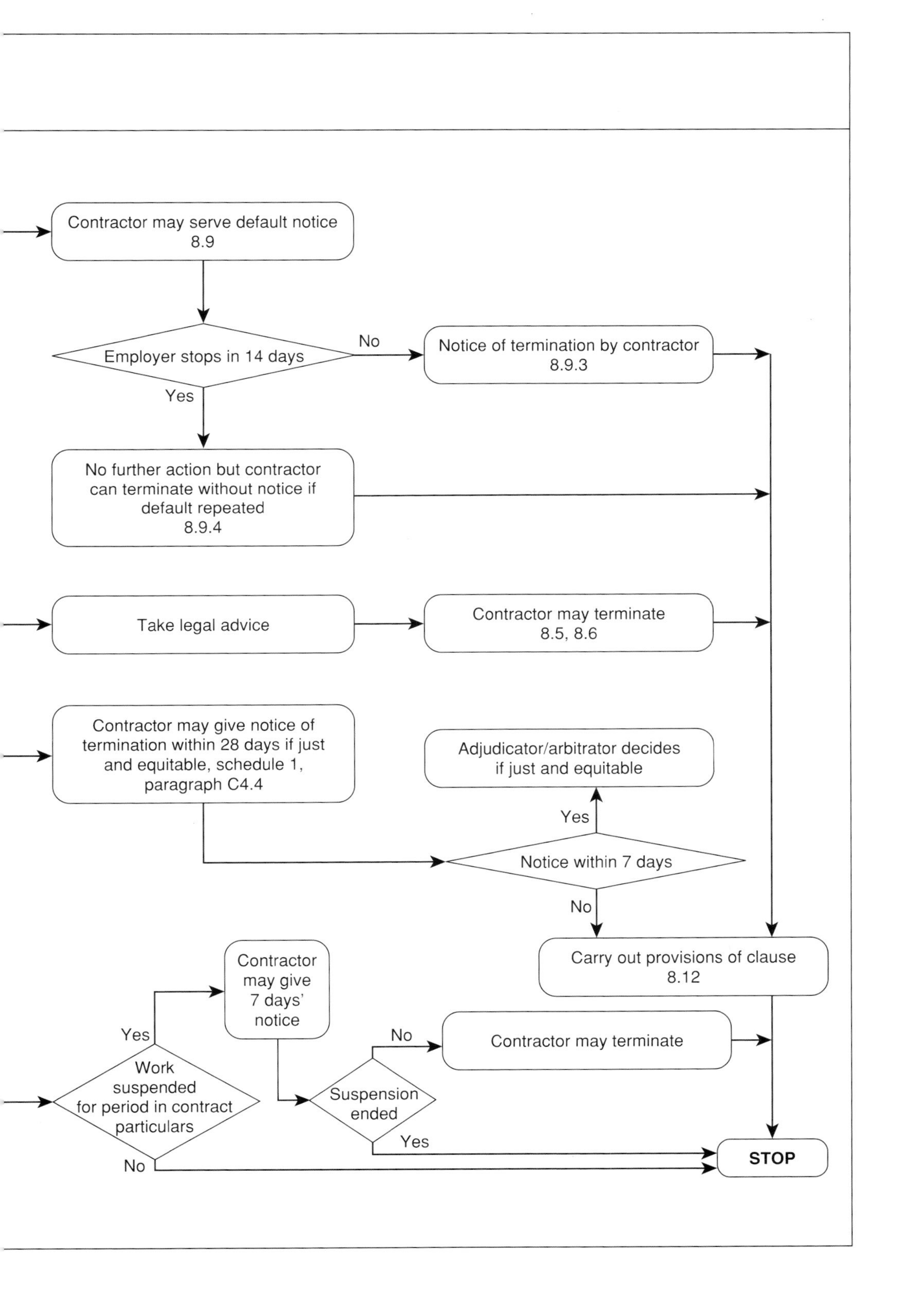
Contractor may serve default notice
8.9
Employer stops in 14 days
No
Yes
Notice of termination by contractor
8.9.3
No further action but contractor can terminate without notice if default repeated
8.9.4
Take legal advice
Contractor may terminate
8.5, 8.6
Contractor may give notice of termination within 28 days if just and equitable, schedule 1, paragraph C4.4
Adjudicator/arbitrator decides if just and equitable
Yes
Notice within 7 days
No
Carry out provisions of clause
8.12
Contractor may give 7 days' notice
Yes
Work suspended for period in contract particulars
No
Suspension ended
No
Yes
Contractor may terminate
STOP

Figure 12.11
Contractor to employer, giving notice of default before termination

SPECIAL DELIVERY

Dear

We hereby give you notice under clause 8.9.1 of the conditions of contract that you are in default in the following respect:

[*Insert details of the default, with dates if appropriate*].

If you continue the default for 14 days from receipt of this notice, or if at any time thereafter you repeat such default (whether previously repeated or not), we may terminate our employment under this contract.

Yours faithfully

Copy: Architect

Figure 12.12
Contractor to employer, giving notice of suspension event before termination

SPECIAL DELIVERY

Dear

We hereby give you notice under clause 8.9.2 of the conditions of contract that the following suspension event has occurred:

[*Insert details of the suspension event with date of commencement*].

If the suspension event is continued for 14 days from receipt of this notice, or if at any time thereafter the suspension event is repeated (for whatever period) whereby the regular progress of the Works is or is likely to be materially affected, we may terminate our employment under this contract.

Yours faithfully

Copy: Architect

default in respect of the matter for 14 days after receipt of the notice, the contractor may within 10 days serve notice by special delivery or recorded or actual delivery to terminate its employment under the contract.

The remarks regarding postage of notices in section 12.1.2 are applicable.

The contractor has the same important power to terminate its employment if the matter is repeated as given to the employer in clause 8.2.3. This means in theory that if the employer, having defaulted once and been given notice, defaults a second time in payment by only one day, or if a suspension event is repeated even for a relatively short period, and the regular progress of the Works is substantially affected as a result, the contractor can serve notice of termination (Figures 12.13 and 12.14). In practice, such a step might be held to be unreasonable or vexatious. But there would be nothing unreasonable in the contractor's giving notice of termination if the payment on a second occasion was a week late, or almost the whole of the Works were suspended for a second time for a week. Most contractors are reluctant to terminate, because it gives them a bad reputation, whatever the reason. So in all likelihood, and prudently, the contractor will send a warning letter in the event of a second default.

Before determining its employment the contractor should consider the five grounds carefully.

The employer does not pay the contractor etc. (clause 8.9.1.1)

This is the employer's responsibility. The architect must make the employer aware, at the beginning of the contract, that prompt payment is vital. The architect's responsibility is to ensure that the employer receives the certificate. This perhaps needs stressing. There are still both architects and contractors who wrongly think the payment period does not begin to run until the contractor presents its copy of the certificate to the employer. Where possible, financial certificates should always be delivered by hand and a receipt obtained. If that is not practical, they should be sent by 'next day' special delivery. The period of time allotted for payment begins to run from the date of issue of the certificate, not from the date of receipt. The date of issue of a certificate has been held to be the date of posting: *Cambs Construction Ltd.* v. *Nottingham Consultants* (1996). The periods are: 14 days for interim payments and the interim payment on practical completion; and 28 days for the final certificate. Termination for failure to pay the amount shown on the final

Figure 12.13
Contractor to employer, terminating employment after repetition of default

SPECIAL DELIVERY

Dear

We refer to the notice of default sent to you on the [*insert date*]. You repeated that default on the [*specify date or dates*] in that you [*describe circumstances briefly*].

Take this as notice that, in accordance with clause 8.9.4 of the conditions of contract, we hereby terminate our employment under the contract without prejudice to any other rights or remedies which we may possess.

We are making arrangements to remove all our temporary buildings, plant, etc. and materials from the Works, and we will write to you again very soon regarding the account which we have started to prepare in accordance with clause 8.12.

Yours faithfully

Copy: Architect

Figure 12.14
Contractor to employer, terminating employment after repetition of specified suspension event

SPECIAL DELIVERY

Dear

We refer to the notice specifying a suspension event sent to you on [*insert date*]. That event was repeated, commencing on [*insert date*], and the effect on regular progress of the Works was [*describe effect briefly*].

Take this as notice that, in accordance with clause 8.9.4 of the conditions of contract, we hereby terminate our employment under the contract without prejudice to any other rights or remedies which we may possess.

We are making arrangements to remove all our temporary plant etc. and materials from the Works, and we will write to you again very soon regarding the account which we have started to prepare in accordance with clause 8.12.

Yours faithfully

Copy: Architect

certificate would be a singularly pointless gesture; the contractor is more likely to refer the dispute to adjudication.

The employer interferes with or obstructs the issue of any certificate (clause 8.9.1.2)

It is important to note that this ground refers to any certificate, not merely financial certificates. There are other certificates (Table 4.3) that the architect is required to issue which the employer conceivably may try to prevent. It will, of course, be difficult for the contractor to prove that the employer is obstructing the issue of a certificate unless the architect tells the contractor. The architect has a clear duty under the contract to issue certificates. It must be made plain to the employer that the employer who tries to interfere with that duty is in breach. If, despite the warning, the employer absolutely forbids the architect to issue a certificate, the architect is in a difficult position. The architect's duty is then to write and confirm the instructions received, setting out the consequences to the employer (Figure 12.15). The architect has no duty to deliberately inform the contractor, but if the contractor suspects and terminates anyway, the architect will be obliged to reveal the facts in any proceedings which may follow.

The employer fails to comply with the assignment provisions (clause 8.9.1.3)

This refers to the prohibition on assignment in clause 7.1, which states that neither party may assign the contract without the written consent of the other. Assignment of rights and duties can be done effectively only by novation, but if the employer purported to assign the right to the completed building to another it would be a serious matter, for which the contractor could pursue termination.

The employer fails to comply according to the contract with the CDM Regulations (clause 8.9.1.4)

This clause refers to a failure on the part of the employer to comply with clause 3.18, which obliges the employer to ensure that the planning supervisor carries out his or her duties correctly. If, unusually, the contractor is not the principal contractor under Regulation 6(5), the employer must also ensure that the principal contractor also carries out its duties properly. If the contractor notifies any amendment to the health and safety plan, the employer must notify the planning supervisor and the architect immediately. The contractor should not

Figure 12.15
Architect to employer if employer obstructs issue of a certificate

Dear

I confirm that a certificate under clause [*insert clause number*] of the conditions of contract is/was [*delete as appropriate*] due on [*insert date*].

I further confirm that you have instructed that I am not to issue this certificate. I am obliged to take your instructions in this matter, but you place me in considerable difficulty, and I need to consider my position. The contractor is certain to enquire about the certificate and, if it suspects that you have obstructed the issue, it may exercise its rights to terminate its employment by reference to clause 8.9.1.2 of the conditions of contract. There will be serious financial consequences for you.

In the light of the above, I look forward to hearing that you have reconsidered your position.

Yours faithfully

hesitate to threaten termination, because the consequences of failure to comply with the Regulations are serious. Contractors should also remember that an employer may be taking that role once in a lifetime, and should be properly advised on the responsibilities by the architect.

The carrying out of the whole or substantially the whole of the Works is suspended for a continuous period of two months or whatever period is stated in the contract particulars (clause 8.9.2)

If the carrying out of virtually the whole of the Works is suspended for the specified period for either of the reasons set out, the contractor may terminate as described. The first reason relates to the architect issuing instructions regarding the correction of inconsistencies, instructions requiring a variation, or instructions postponing the carrying out of the work. If the contractor is delayed for two months thereby, it will be in very serious trouble. For any period up to two months, it would be entitled of course to put together a claim for loss and/or expense under clause 4.17. For any contractor handling this value of work, a two-month delay could be disastrous. This clause quite reasonably gives the contractor the option of termination if it foresees no quick end to the suspension and it feels unable to afford to keep the site open. The clause unnecessarily emphasises that the delay must not be due to the contractor's own negligence or default.

The second reason is any impediment or default of the employer, the architect, the quantity surveyor or the employer's persons. The very broad, catch-all clause replaces a series of clauses in IFC 98 which included a failure of the architect to provide information on time, work or materials which the employer was going to supply and the employer's failure to provide access. The new clause renders them all redundant, because they are all examples of default by the employer or persons engaged or authorised by the employer.

12.2.3 Grounds (clause 8.10): insolvency

The grounds for termination under this clause are the same as for the contractor's insolvency under clause 8.5 (see section 12.1.3). Termination is not automatic, and notice must be served by special or recorded or actual delivery. It takes effect on receipt by the employer (clause 8.2.2). It is highly unlikely that the contractor would wish to continue and take its chance of being paid. After the employer becomes insolvent, and before the notice of termination takes effect, the contractor's obligation to proceed and complete the Works is suspended. This is to avoid the silly situation which would otherwise exist during this, probably brief, period when the contractor would be legally obliged to continue until it could terminate its employment. Under clause 8.10.2 the contractor must be given notice in writing if the employer makes a proposal, calls a meeting or becomes subject to proceedings or appointment in regard to any of the insolvency matters listed in clause 8.1.

12.2.4 Grounds (clause 8.11): neutral causes

The grounds for termination under this clause have already been covered in section 12.1.5. If the contractor wishes to terminate, however, there is a proviso (clause 8.11.2) to the effect that the contractor is not entitled to give notice if the loss or damage due to specified perils is caused by the contractor's own negligence or default or of its servants or agents or anyone employed on the Works other than the employer or a person engaged by the employer or by a local authority or statutory undertaking carrying out work solely in accordance with statutory obligations. This proviso is only stating expressly what must be implied – that the contractor must not be able to profit by its own default. (Figure 12.16 is a suitable letter.)

12.2.5 Grounds (paragraph C.4.4 of schedule 1 and clause 6.10.2.2): insurance risks and terrorism cover

The grounds and procedure for termination under this clause are exactly the same as for the employer (see section 12.1.6).

12.2.6 Consequences (clause 8.12)

This clause lays down the procedure to be followed after termination under clauses 6.10.2.2, 8.9, 8.10, 8.11 and paragraph C.4.4 of schedule 1. It is refreshing to see that JCT has tidied up and simplified these provisions. They are a marked improvement over the equivalent provisions in IFC 98. The following should be noted:

- There is no longer any provision requiring that the contractor must remove from site all its temporary buildings, plant, equipment,

Figure 12.16
Contractor to employer, terminating employment after repetition of default

SPECIAL DELIVERY

Dear

In accordance with clause 8.9.3 of the conditions of contract, we hereby determine our employment under the contract, because [*insert appropriate details*].

We are making arrangements to remove all our temporary buildings, plant, etc. and materials from the Works, and we will write to you again very soon regarding the account which we have started to prepare in accordance with clause 8.12.

Yours faithfully

Copy: Architect

etc., but no doubt a contractor in this position would not need any encouragement to do so

- There is no express provision for the contractor to give up possession of the site, which is a pity, but the point must be academic, because if the contractor has removed all its plant etc. already, it can hardly claim to be in possession
- The contractor's liability to insure the Works ends on termination, and the architect must immediately remind the employer to insure
- Clause 8.12.1 makes clear that any terms of the contract which require further payment or release of retention will cease to apply
- Without wasting any time the contractor must prepare an account setting out the following:
  - the total value of the work done at the date of termination together with any other amounts due under the terms of the contract (e.g. direct loss and/or expense under clause 4.17)
  - the cost of materials properly ordered for the Works for which the contractor has paid or is legally bound to pay (i.e. because a contract has been entered into). Materials 'properly' ordered are those which it is reasonable that the contractor has ordered at the time of termination. Regard must be had to the suitability of the materials and also to the delivery period. Materials paid for become the property of the employer
  - any direct loss and/or damage caused to the contractor by the termination. This is potentially the most damaging clause to the employer, depending on the value of the contract remaining incomplete. Quite rightly, and in accordance with normal contract principles, the contractor is to be paid the profit it would have expected to have made if the contract had run its course. Such loss and/or expense is to be included only if the termination is under clause 8.9 (employer's default), 8.10 (employer's insolvency) or 8.11.1.3 (where the loss or damages resulted from the negligence or default of the employer or persons authorised or acting as agents of the employer)
  - less any sums previously paid to the contractor.

Clause 8.12.4 states that the employer must pay the amount properly due within 28 days of its submission by the contractor and without any deduction of retention. The employer will clearly require the architect and the quantity surveyor (if appointed) to check the account, and the employer's obligation is merely to pay the amount properly due. If it is intended to pay less than the balance shown on the contractor's account, the employer should issue a written notice to the contractor to that effect within five days of receiving the account. The notice should be just as detailed as a notice issued under clause 4.8.2. Clause 8.12.4 makes no express reference to this, but the requirements of the Housing Grants, Construction and Regeneration Act 1996 appear to apply. If no such notice is served, it appears that the employer is still entitled to serve a set-off notice complying with clause 4.8.3, no later than five days before the expiry of the 28 days. However, it is clear that the account submitted by the contractor is not the 'amount due'. Therefore, even if no notices were served, the employer would not be obliged to pay the amount in the contractor's submission, if it could be demonstrated that a different amount was in fact due.

12.2.7 Consequences (paragraph C.4.4 of schedule 1 and clause 6.10.2.2)

After either party has terminated the employment of the contractor under paragraph C.4.4 of schedule 1 or clause 6.10.2.2, all the provisions of clause 8.12 (see section 12.2.6) apply, except that the contractor is not entitled to any direct loss and/or damage caused by the termination. This is to reflect the fact that these consequences refer to termination for causes beyond the control of the parties.

The architect is not required to issue a final or any other certificate after termination for any reason.

12.3 *Summary*

Grounds for termination by the employer

- The contractor stops work without good reason
- The contractor fails to proceed diligently
- The contractor fails to comply with the architect's notice, and serious consequences follow
- The contractor fails to comply with certain clauses
- The contractor fails to comply with the CDM Regulations
- The contractor becomes insolvent
- The contractor is corrupt.

Grounds for termination by the contractor

- The employer does not pay on time
- The employer obstructs a certificate
- The employer assigns without consent

- The employer fails to comply with the CDM Regulations
- The work is stopped for a stipulated period because of certain instructions or acts of prevention or impediment on the part of the employer, architect, quantity surveyor or employer's persons
- The employer becomes insolvent.

Grounds for termination by either party

- Terrorism cover ceases to be obtainable
- Damage is caused to work to an existing building by specified perils, and it is fair to terminate
- The work is stopped for a specified period because of *force majeure*, architect's instructions regarding inconsistencies, variations or postponement, loss or damage due to specified perils, civil commotion or terrorism, statutory powers exercised by UK government.

CHAPTER THIRTEEN
CONTRACTOR'S DESIGNED PORTION (CDP)

13.1 *General*

Architects and contractors commonly assumed that IFC 98 gave design responsibility to the contractor. Nothing could be further from the truth. The contractor had no design responsibility under IFC 98, and the position is exactly the same under IC.

It has long been established that the overall responsibility for design rests with the architect, and the architect can avoid this responsibility only by obtaining the employer's express consent to assigning the responsibility to another: *Moresk Cleaners Ltd* v. *Hicks* (1966). A named sub-contractor may be given design responsibility, but the contractor is not responsible for the sub-contractor's failure in this respect (see schedule 2, paragraph 11.1), and the architect is still ultimately liable unless the employer has expressly consented to the transfer of liability to the named sub-contractor. In the event of a named sub-contractor having design responsibility, it is essential that the warranty ICSub/NAM/E is completed. Indeed, ICSub/NAM/E should be completed for each named sub-contractor in any event.

The Intermediate Building Contract with Contractor's Design 2005 (ICD), as the name suggests, incorporates extensive provisions to enable the contractor to be given design responsibility for specific items. This must not be confused with any design responsibility of a named sub-contractor.

In essence, the CDP provisions are a design and build contract in miniature and share many of the features of the DB contract, albeit in briefer terms.

13.2 *Documents*

Details of the CDP are to be inserted in the second recital. A footnote, somewhat superfluously, advises that a separate sheet may be used if the space is not sufficient to include all the items. If indeed a separate sheet is required, the employer ought to seriously consider whether the DB contract would not be more suited. The employer is warned, in bold, not to include any part of named sub-contract work. To ignore that warning is the recipe for confusion – at the very least.

The sixth and seventh recitals indicate the process whereby the employer provides a set of Employer's Requirements with the other documents included in the invitation to tender and, in response, the contractor provides a set of Contractor's Proposals showing the design and construction together with an analysis of the portion of the contract sum that relates to the CDP. Details of these documents are to be inserted in the contract particulars for identification purposes. When the contract is executed, and these documents are signed by the parties, they become part of the contract documents.

A question that continually arises in regard to the DB contract and to the CDP portion is whether, in case of conflict, the Employer's Requirements or the Contractor's Proposals take precedence. It was open to JCT to unequivocally state the position, but it is strangely reluctant to do so. Indeed, it perpetuates the confusion by retaining, in the seventh recital, the rather ambiguous statement that the employer has examined the Contractor's Proposals and 'is satisfied that they appear to meet the Employer's Requirements'. That seems to be the equivalent of saying 'I am reasonably certain that I am not sure'. Fortunately, the recitals are only to be called in aid to construing the contract when the operative part of the contract is ambiguous: *Rutter* v. *Charles Sharpe & Co Ltd* (1979). A careful reading of other clauses in the contract makes clear that the Employer's Requirements take precedence (see clauses 2.1, 3.11.3 and 5.2.3). In any event, the whole philosophy is that the contractor is required to provide what the employer has set out in the Employer's Requirements, not that the employer is obliged to accept what the contractor offers in the Contractor's Proposals.

13.3 The contractor's obligations

The contractor's obligations are set out in an enlarged clause 2.1. Referring to the CDP, it requires the contractor to:

- Complete the design for the CDP, including the selection of specifications for materials, goods and workmanship to the extent that they are not stated in the Employer's Requirements or the

Contractor's Proposals. It is clear from this that compliance with the Employer's Requirements is the primary objective
- Comply with the architect's instruction about the integration of the CDP work with the rest of the Works, subject to the contractor's right to object under clause 3.8.2
- Comply with the relevant parts of the CDM Regulations, particularly as they affect the designer.

Integration with the rest of the Works is a fruitful area for claims. The contractor will often contend that the architect's instructions for integration unavoidably result in additional work and, therefore, cost. It is easy to confuse the situation. The principles, however, are straightforward, although application to particular circumstances may need some thought. There are four basic situations:

- If the invitation to tender is supported by clear documents showing the rest of the design and especially any likely interfaces with CDP work, it is a matter for the contractor to allow in the Contractor's Proposals for the proper integration of the CDP with the rest of the design. If the rest of the design, so far as it affects the CDP, remains unchanged, and if the architect does not instruct a variation under clause 3.11.3 requiring an alteration in the Employer's Requirements, the contractor can have no claim for any additional cost.
- If the invitation to tender is not supported by sufficient information to enable the contractor to properly design the interface between the CDP and other work, and the contract documents are executed without the ambiguity being clarified, the contractor, in principle, has a claim for any additional cost resulting from the architect's directions on integration.
- If the architect subsequently issues instructions either regarding the rest of the work which affects the CDP or regarding the CDP through the Employer's Requirements, the architect must issue directions on integration and the contractor has a claim for additional cost.
- If the contractor is obliged to alter the CDP in order to correct its own error, the contractor must bear those costs itself even though the architect will probably have to issue some directions about the integration of the corrected CDP.

Where materials, goods and workmanship are neither stated to be to the architect's satisfaction nor described in the contract documents, they are to be of a standard appropriate to the CDP.

Under clause 2.10.2 the contractor is obliged to provide the architect with two copies of design documents reasonably necessary to explain or amplify the Contractor's Proposals. The architect is entitled to request any related calculations or information. The contractor must also provide two copies of any levels or other setting-out information used for the CDP. Clause 2.10.3 suggests that procedures for submission of design information from the contractor should be set out in the contract. The clause refers to the information being provided 'as and when necessary from time to time in accordance with any design submission procedures set out in the Contract Documents'. The contract is silent about the position if no such procedure is set out. The timing of the submissions is covered by the clause. It is the mechanism of the submission which would be missing: for example, how long the architect needed to consider the submission, and whether some graded system of approvals is intended. What is clear is that the contractor may not proceed with any details until they have been submitted. That statement implies that, in the absence of any stated contractual procedure, the contractor is entitled to continue to execute the work as soon as the submission takes place. There is no requirement in the clause itself for any approval or comment process.

Inconsistencies are dealt with in clause 2.13. Clause 2.13.4 adds the CDP documents onto the list of other documents in the general clause. Clauses 2.13.3.2 and 2.13.4 deal with inconsistencies within the individual CDP documents themselves in much the same way as the DB contract. Where there is an inconsistency in the Employer's Requirements which is dealt with in the Contractor's Proposals and is consistent with statutory requirements, what is in the Contractor's Proposals will take precedence. However, if the item is not dealt with in accordance with statutory requirements, the contractor, on discovery, must make proposals for amendments.

Clause 2.14 purports to deal with the resultant financial situation. However, it does not expressly cover clauses 2.13.3.2 and 2.13.4, and the user is left to assume that no adjustment to the contract sum is permissible in those instances, if for no other reason than that no adjustment to the contract sum may take place unless expressly authorised by the contract (clause 4.2). Clause 2.14 confines itself to instructions under clause 2.13.1 which constitute a variation, which are to be valued under clause 5. It prohibits any addition to (but does not preclude a deduction from) the contract sum or any extension of time resulting from instructions issued to correct errors or inconsistencies in the CDP documents or necessary variations excepting the Employer's Requirements. It also prohibits any additions to the contract sum resulting from delay or suspension which itself was caused by the contractor's failure to comply with regulation 13 of the CDM Regulations or to provide at the right time any design documents or other information as required by clause 2.10.3 or as requested by the architect in writing.

Clause 2.15 deals with divergences from statutory requirements, and this clause has been expanded to incorporate references to the CDP. If one or more of the CDP documents is involved, the contractor must submit its proposals for removing the divergence. It is to be noted that clause 2.15.2 requires the architect to issue instructions within 14 days of receipt of this submission. The contractor is to comply with the instruction at no cost to the employer unless the divergence is a result of a change in statutory requirements after the base date. In that instance, the amendment to the CDP is to be treated as a variation.

Before practical completion, the contractor is required to provide the employer with whatever drawings and other information is specified in the contract documents or, even if not so specified, such information as the employer may reasonably require which shows the CDP as built including the maintenance and operation of the CDP and any included installations (clause 2.32). This is expressly stated not to affect the contractor's obligations to provide information for the health and safety file under clause 3.18.

13.4 Liability

Clause 2.33 stipulates that the employer is to have an irrevocable, royalty-free and non-exclusive licence to use the CDP documents in various ways, but in the usual way excluding the right to reproduce the designs for an extension to the Works. Irrevocable means exactly that. Once the contract is executed by the parties, the licence cannot be revoked for any reason. However, the clause is subject to the overriding proviso that all monies due and payable to the contractor have been paid. That will normally mean that the monies shown due on all certificates have been paid. It is obvious that, although the employer can use the designs so long as payment is properly made to the contractor, the licence will cease to be effective if proper payment stops. It seems that the employer could be in a serious situation if payment is stopped without good reason, because the reproduction of the design which, until then, was lawful, will suddenly become an infringement of the contractor's copyright. In that situation, the employer would be the author of its own misfortune. Quite rightly, the contractor is not to be liable for the misuse of any of its design documents. Anyone called upon to produce a design for another person according to a specific brief will know that the finished design is seldom used precisely as the original intention. That is because ideas change. There is nothing wrong with that – quite the contrary – but the contractor cannot be held liable for any change of use. The clause is virtually superfluous, because the contractor could not be held liable in any event for the result of employer misuse unless the likelihood of misuse is so notorious as to be reasonably foreseeable: *Introvigne* v. *Commonwealth of Australia* (1980).

Clause 2.34 sets out the design liabilities and limitations. There are some interesting features. For what it designed or completed, the contractor would normally have a fitness for purpose liability (*Viking Grain Storage Ltd* v. *TH White Installations Ltd* (1985)), but that is modified by clause 2.34.1 to the same standard as that of an architect: that is, reasonable skill and care. Therefore, like an architect, the contractor does not guarantee the result of the design, but only that reasonable skill and care was taken in its production.

The contractor is also liable under the Defective Premises Act 1972 (in Northern Ireland the Defective Premises (Northern Ireland) Order 1975) if the CDP involves dwellings. If dwellings are not involved, clause 2.34.3 provides an option for the parties to restrict the contractor's liability for loss of use, loss of profit and other consequential loss. An appropriate sum is to be inserted in the contract particulars. It is not clear why the employer would want to agree to this, other than perhaps to reduce the contractor's tender. It is doubtful whether the contractor would be liable for such losses in any event unless the likelihood was expressly made known to the contractor before the contract was executed: *Hadley* v. *Baxendale* (1854).

An interesting new clause deals with existing designs (clause 2.34.4). It is expressly stated that the contractor is not responsible for what is in the Employer's Requirements, nor for verifying whether any design is adequate. This clause is in response to *Co-operative Insurance Society Ltd* v. *Henry Boot Scotland Ltd* (2002), where the court considered an earlier edition of the Contractor's Designed Portion Supplement and held that the contractor had a duty to check any design it was given and to make sure that it worked. It had previously been assumed that the words '. . . the Contractor shall complete the design . . .' and later that '. . . the design . . . is comprised in . . . what the Contractor is to complete . . .' were intended to make the contractor liable only for any additional design. It was thought that any defective design included in the Employer's Requirements was a matter for the employer alone. The insertion of this clause appears to restore that position.

In acknowledgement of that, clause 2.34.5 provides that if any inadequacy is found in the Employer's Requirements, and if the Contractor's Proposals have not corrected it, the Employer's Requirements must be corrected, and the correction is to be treated as a variation. This clearly includes any defective design. The provision is said to be subject to clause 2.15, which of course deals with divergences from statutory requirements (see above).

There is the usual prohibition on the contractor, without the architect's written consent, sub-letting the design of the CDP except in accordance with clause 3.6, which stipulates certain requirements of any sub-contract. It is probably worth noting that, provided the contractor does comply with clause 3.6, the architect's consent is not required.

There are a few minor amendments to the contract resulting from the CDP. In clause 2.3 the contractor is not to be reimbursed any fees or charges which it pays solely relating to the CDP work. Where, under clause 2.16, emergency action is necessary because of a divergence between statutory requirements and CDP documents, it is not to be treated as a variation. The relevant event clause 2.20.11 is expanded to include any persons engaged in the preparation of a design for the CDP as affected by strikes, lock-outs or local combination of workmen.

13.5 *Variations*

It is made clear in clause 3.11.3 that an instruction requiring a variation to the CDP can be issued only in respect of the Employer's Requirements. The employer cannot issue an instruction directly about the design in the Contractor's Proposals. Therefore it is for the architect to instruct a change to the Employer's Requirements, to which the contractor responds by altering its design. Clauses 5.2.3 and 5.2.4 confine instructions about the expenditure of provisional sums and work as approximate quantities to sums or quantities in the Employer's Requirements. There is no provision to instruct the expenditure of provisional sums or work as approximate quantities in the Contractor's Proposals. These are further reasons why the Employer's Requirements take precedence.

There is a very important qualification to the contractor's obligation to comply with the architect's directions or instructions. It is contained in clause 3.8.2. It states that, if the contractor's opinion is that compliance will 'injuriously' affect the design of the CDP, the contractor must specify that effect by written notice to the architect within seven days of receipt of the direction or the instruction. Unless and until confirmed by the architect, the contractor need not comply. Presumably, having been 'put on notice' by the contractor, the architect will not lightly require compliance without carefully checking the grounds of the contractor's objection.

Valuation of variations in CDP work is to be carried out in accordance with clause 5.7. This is a very short clause, but an important one. The base document is the CDP analysis. The valuation process is modelled on the valuation rules applicable to general variations, and clause 5.7.4 provides that clauses 5.3.3.2, 5.3.3.3, 5.4 and 5.5 are to apply so far as they are relevant. Whether and to what extent they are relevant will be decided by the architect or the quantity surveyor, as the case may be, doing the valuation.

In general, the valuation of varied work is to be consistent with the value of work of similar character in the CDP analysis, but due allowance must be made for change in the conditions under which the work is carried out. If there is significant change in quantity, allowance must be made for that also. What is significant, ideally will be a matter for the quantity surveyor to decide based on experience of when economies of scale begin to take effect. If there is no work of a similar character to the variation in the CDP analysis, a fair valuation is to be made in the usual way. Where CDP work is omitted, it is to be valued in accordance with the values in the CDP analysis.

In common with the DB contract, allowance must be made in the valuation of CDP work for addition or omission of design work. How this is to be done is not stated. It is therefore important that the contractor includes among its prices in the CDP analysis something for design. The easiest way to do that is probably as an hourly rate or rates. If no method of calculating the cost of design is included, the contractor may well be deprived of substantial additional costs where extra CDP work is instructed. This is particularly the case where the design work is aborted so that the contractor has produced a design in answer to an instruction to vary some CDP work and the instruction is then withdrawn. If design is priced separately, the contractor will be entitled to payment at that rate for the design work done, but not used. If there is no way of separating the design work, the contractor may be paid nothing at all, on the basis that if the rate for design is part of the rate for carrying out the work, there can be no payment for design if the work is not carried out. At best, the contractor may be looking at a nominal payment. Of course, the contractor must be scrupulous in keeping detailed time sheets for the design work to substantiate any claim for payment.

13.6 Insurance

Clause 6.15 requires the contractor to take out professional indemnity insurance. The type and amount are to be in accordance with what is stated in the contract particulars. The details must be very carefully completed, because failure to do so may result in the insurance not being required or being in inappropriate terms. The employer should take advice from an experienced broker before this part is completed. The period for insurance runs from the date of practical completion for either 6 or 12 years. Twelve years should be chosen to match the limitation period if the contract is executed as a deed. The insurance is to be taken out forthwith after the contract is entered into unless it has already been taken out. Usually, the design will be done by others, and they will have ongoing professional indemnity insurance. The contractor will probably try to put forward its sub-contractors' insurances to satisfy this clause. On a strict reading of the clause, such sub-contract insurance would not be acceptable. The contractor should be the insured.

The contractor must maintain the insurance until the expiry of the period stated in the contract particulars. There is a proviso to that, which will be welcome to the contractor and with which most construction professionals will be familiar. The proviso is that the insurance remains available at commercially reasonable rates. This is a fairly wide proviso. What is commercially reasonable will depend on several factors, not least the contractor's own financial circumstances. In order to decide whether the rate is commercially reasonable it is probably simply necessary for the contractor to ask itself whether it makes commercial sense to maintain the insurance at the particular rate quoted. Unless the contractor is carrying out CDP work on a regular basis, it is doubtful whether it would make commercial sense except at a fairly nominal amount. However, all the circumstances must be taken into account.

Clause 6.16 attempts to deal with the situation if the insurance ceases to be available at commercially reasonable rates. Although the situation is viewed as a factual occurrence, it is clear that it will depend on the subjective judgment of the contractor. It is not obvious how an employer could argue that a rate was commercially reasonable if the contractor argued to the contrary. In the event that the insurance is terminated for such reason, the contractor must immediately give notice to the employer, with the intention that the employer and the contractor will discuss how best to protect their respective positions. The employer is obviously in a very exposed position. Probably, it will be difficult to contend that the contractor is in breach if it puts forward the unreasonable commercial rate ground. The employer cannot oblige the contractor to continue the insurance. Until grounds for a claim arise, the employer will have suffered no loss as a result of the lack of insurance. Once the claim arises, it will be too late. The contractor, if a limited company, may simply liquidate. This is a real problem to which there is no easy answer, short of the employer taking out the insurance.

An important clause (6.15.3) requires the contractor to provide the employer or the architect with documentary evidence that the insurance has been effected or is being maintained.

13.7 Summary

- The contractor has no design liability under the IC contract
- CDP is like a mini design and build contract set in IC
- Unless there is an inconsistency within them which is dealt with by the Contractor's Proposals, the Employer's Requirements take precedence
- The contractor must complete the design for the CDP work, but it is not responsible for checking any design in the Employer's Requirements
- The architect must issue any necessary directions for integration of the CDP with the rest of the Works
- The contractor has the same liability for design as an architect
- Valuation of variations of CDP work must include something for design
- The contractor must take out professional indemnity insurance if so stated in the contract particulars.

CHAPTER FOURTEEN
DISPUTE RESOLUTION PROCEDURES

14.1 General

The contract provides for four systems of dispute resolution.

The first system, about which little need be said, is mediation. This is dealt with by clause 9.1. The clause is very brief and simply states that, by agreement, the parties may choose to resolve any dispute or difference arising under the contract through the medium of mediation. A footnote refers readers to the Guide. It is unclear why this clause has been included in the contract at all. It was not in the IFC 98 edition except in the form of a footnote. The key to the redundant nature of this clause is in the phrase 'The Parties may by agreement . . .'. The parties, of course, may do virtually anything by agreement. They can agree to scrap the whole contract and sign a different one if they are both of one mind on the matter. One assumes that this clause was inserted purely to remind the less sophisticated users of the form that mediation is a possibility. Why the draftsman of the contract did not also refer to the possibility of conciliation or negotiation is not clear. In general, there is little point in including as terms of a contract anything to be agreed. The whole point of a written contract is that it is evidence of what the parties have already agreed. To have a clause that effectively states 'We may agree to do something else' is a waste of space. It is to be hoped that future editions of the contract consign this clause either back to a footnote or, better still, to the Guide where it belongs.

In 1996 the Housing Grants, Construction and Regeneration Act (commonly called the Construction Act) (the Act) was enacted (in Northern Ireland Part II of the Act is virtually identical to the Construction Contracts (Northern Ireland) Order 1997). Section 108 of the Act expressly introduces a contractual system of adjudication to construction contracts. Excluded from the operation of the Act are contracts relating to work on dwellings occupied or intended to be occupied by one of the parties to the contract. IC and ICD, in common with other standard forms, incorporate the requirements of the Act. Therefore all construction Works carried out under these forms are subject to adjudication even if they comprise work to a dwelling house. In essence section 108 provides that:

- A party to a construction contract has the right to refer a dispute under the contract to adjudication
- Under the contract:
  - A party can give notice of intention to refer to adjudication at any time
  - An adjudicator should be appointed and the dispute referred within seven days of the notice of intention
  - The adjudicator must make a decision in 28 days or whatever period the parties agree
  - The period for decision can be extended by 14 days if the referring party agrees
  - The adjudicator must act impartially
  - The adjudicator may use his or her initiative in finding facts or law
  - The adjudicator's decision is binding until the dispute is settled by legal proceedings, arbitration or agreement
  - The adjudicator is not liable for anything done or omitted in carrying out the functions unless in bad faith
- If the contract does not comply with the Act, the Scheme for construction contracts (England and Wales) Regulations 1998 (the Scheme) will apply.

The right to refer to adjudication 'at any time' means that adjudication can be commenced even if legal proceedings (and presumably arbitration) are in progress about the same dispute: *Herschel Engineering Ltd* v. *Breen Properties Ltd* (2000). It has also been held that adjudication can be sought even if repudiation of the contract has taken place. A dispute may be referred to adjudication, and the adjudicator may give a decision, even after the expiry of the contractual limitation period: *Connex South Eastern Ltd* v. *MJ Building Services plc* (2005). Of course, in such a case, the referring party runs the risk that the respondent will use the limitation period defence – in which case the claim will normally fail.

A clause in the contract appears to contravene the requirements of the Act that a party has the right to refer to adjudication 'at any time'. Paragraph C.4.4.1 of schedule 1 states that either party may 'within seven days of receiving [a termination notice] (*but not thereafter*) invoke the dispute resolution procedures that apply . . .' (emphasis added). This is a clear attempt to restrict the period during which a referral may take place. It would almost certainly be ineffective against the adjudication clause.

Adjudication is dealt with in article 7 and clause 9.2.

Adjudication is what might be termed a 'temporarily binding solution'. In the vast majority of cases, anecdotal evidence suggests that the parties accept the adjudicator's decision and do not take the matter further. Even where there are challenges through the courts against the enforcement of an adjudicator's decision, the challenge is concerned with matters such as the adjudicator's jurisdiction or whether the adjudicator complied with the requirements of natural justice, not whether the adjudicator's decision was correct. The courts cannot interfere with the adjudicator's decision, no matter how obviously wrong, provided that the adjudicator has answered the questions posed by the referring party: *Bouygues United Kingdom Ltd* v. *Dahl-Jensen United Kingdom Ltd* (2000).

The parties must comply with the adjudicator's decision, following which, if they are not satisfied, either party may instigate proceedings through the stipulated system of obtaining a final decision. It is important to remember that, in doing so, the parties are not appealing against the decision of the adjudicator, and the arbitrator or court will ignore the adjudicator's previous decision in arriving at an award or judgment respectively: *City Inn Ltd* v. *Shepherd Construction Ltd* (2002).

If the parties wish to have a final and binding decision rather than submit a dispute to adjudication, they have a choice between arbitration and legal proceedings. Note that, unlike the position under previous contracts up to and including IFC 98, legal proceedings will apply unless the contractor particulars are completed to show that arbitration is to be the procedure. One suspects that the decision to default to legal proceedings rather than arbitration was taken after significant lobbying by solicitors, who traditionally feel at home with court proceedings and, with a few exceptions, unhappy with arbitration. The advantages of arbitration are often said to be:

- *Speed* – Much depends on the arbitrator, but a good arbitrator should dispose of most cases in months, not years
- *Privacy* – Only the parties and the arbitrator are privy to the details of the dispute and the award
- *The parties decide* – The parties can decide the timescales, the procedure and the location of any hearing
- *Expense* – Theoretically, it should be more expensive than litigation, because the parties (usually the losing party) have to pay for

the arbitrator and the hire of a room, but in practice the speed and technical expertise of the arbitrator usually keep costs down
- *Technical expertise of the arbitrator* – The fact that the arbitrator understands construction should shorten the time schedule and possibly avoid the need for expert witnesses if the parties agree
- *Appeal* – The award is final, because the courts are loath to consider any appeal.

Possible disadvantages are:

- In theory, it is more expensive because the parties (usually the losing party) pay the cost of the arbitrator and the hire of a room for the hearing
- If the arbitrator is poor, the process may be slow and expensive
- The arbitrator may not be an expert on the law, which may be a major part of the dispute
- Parties who are in dispute may find it difficult to agree about anything. Therefore the arbitrator may be appointed by the appointing body, and the procedure, the timing and the location of the hearing room may be decided by the appointing body and the arbitrator as the case may be, with the result that neither party is satisfied.

The advocates of legal proceedings list the advantages as follows:

- The judge should be an expert on the law
- The Civil Procedure Rules require judges to manage their caseloads and encourage pre-action settlement through use of the Pre-Action Protocol
- Cases can reach trial quickly
- The claimant can join several defendants into the proceedings to allow interlocking matters and defendants to be decided
- Costs of judge and courtroom are minimal
- A dissatisfied party can appeal to a higher court.

The disadvantages of legal proceedings are said to be:

- Even specialist judges know relatively little about the details of construction work
- Parties cannot choose the judge, who may not be very good
- Costs will be added because expert witnesses or a court-appointed expert witness will be needed to assist the judge
- Cases often take years to resolve

- Lengthy timescale and complex processes may result in high costs
- Appeals may result in an unacceptable level of costs.

Arbitration is probably still the most satisfactory procedure for the resolution of construction disputes, and employers would be advised to complete the contract particulars accordingly. Where the parties have agreed that the method of binding dispute resolution will be arbitration, a party who attempts to use legal proceedings instead will fail in a costly way if the other party calls in aid section 9 of the Arbitration Act 1996. Section 9 requires the court to grant a stay (postponement) of legal proceedings until the arbitration is concluded unless the arbitration is null, void, inoperable or incapable of being performed. The court has no discretion about the matter, and the successful party will claim its costs. The result is that the party intent on legal proceedings will not only have to revert to arbitration, but that it will also have to pay the other party's legal costs in opposing the legal proceedings.

Arbitration is dealt with by article 8 and clauses 9.3 to 9.8. Legal proceedings are dealt with by article 9.

14.2 *Adjudication*

14.2.1 The contract provisions

Article 7 provides that, if any dispute or difference arises under the contract, either party may refer it to adjudication in accordance with clause 9.2. The parties are of course the employer and the contractor. Although in most instances it will be the contractor that initiates adjudication, there is nothing to stop the employer from doing so. For example, the employer may seek adjudication if the architect over-certifies or makes an extension of time that seems excessive.

The architect is not a party to the contract, and therefore cannot be the respondent to an adjudication. Architects can obviously act as witnesses, but they have no duty to run an adjudication on behalf of the employer. Acting in an adjudication usually calls for some degree of skill and experience which most construction professionals, acting in the normal course of their professions, will not readily acquire. Where the dispute is other than very straightforward, or where one party has retained the services of a legal representative, the other party is well advised to do likewise.

Note that only disputes arising 'under' the contract may be referred. This can be contrasted with arbitration, where the disputes are described as 'of any kind whatsoever arising out of or in connection with this Contract'. This phrase has been considered (*Ashville Investments Ltd* v. *Elmer Contractors Ltd* (1987)) and held to be very broad in meaning. Thus, for example, an adjudicator has no power to consider formal settlement agreements about various matters made by the parties in connection with the contract of which the adjudication clause forms part: *Shepherd Construction* v. *Mecright Ltd* (2000).

Parties often believe that reference to adjudication is necessary before seeking arbitration or legal proceedings as the case may be. That is certainly not the case. Use of the word 'may' makes clear that adjudication is optional. Nevertheless, it is rapidly replacing arbitration as the standard dispute resolution process, albeit that it is rather rough and ready. Unfortunately, it is often used for complex disputes involving large amounts of money for which it is not suited. Adjudications involving £1 million or more with time extended by agreement to two or three months are a travesty of what the Act intended.

Clause 9.2 is much shorter than the comparable clause in IFC 98. That is because the procedure set out in IFC 98 has been abandoned in favour of the procedure in the Scheme. This is an eminently sensible approach, because the Scheme is a comprehensive set of rules especially drafted to comply with the Act. Currently, there is an unwarranted proliferation of procedures.

Use of the Scheme is made subject to certain provisos:

- The adjudicator and nominating body are to be those stated in the contract particulars
- If the dispute concerns whether an instruction issued under clause 3.15 (work not in accordance with the contract) is reasonable, the adjudicator, if practicable, must be someone with appropriate expertise and experience. If not, the adjudicator must appoint an independent expert with appropriate expertise and experience to give advice and to report in writing whether the instruction is reasonable in all the circumstances.

It is unclear why clause 3.15 should have been singled out, because it is perfectly proper to argue that the adjudicator must always have relevant expertise and experience or seek expert assistance.

14.2.2 The Scheme: notice of adjudication

Paragraph 1 provides that any party to a construction contract may give to all the other parties a written notice of an intention to refer a dispute to adjudication. The notice must describe the dispute and who the parties involved are. It must give details of the time and location, the redress sought, and the names and addresses of the parties to the contract.

The notice is the trigger for the adjudication process, and it is also one of the most important documents. Great care must be taken in its preparation, because the dispute that the adjudicator is entitled to consider is the dispute identified in the notice of adjudication: *McAlpine PPS Pipeline Systems Joint Venture* v. *Transco plc* (2004). The dispute cannot be broadened later by the referring party, although it can be elaborated and more detail provided: *Ken Griffin and John Tomlinson* v. *Midas Homes Ltd* (2000). In that case, the court set out the purposes of the notice as:

> '. . . first, to inform the other party of what the dispute is; secondly, to inform those who may be responsible for making the appointment of an adjudicator, so that the correct adjudicator can be selected; and finally, of course, to define the dispute of which the party is informed, to specify precisely the redress sought, and the party exercising the statutory right and the party against whom a decision may be made so that the adjudicator knows the ambit of his jurisdiction.'

For example, if the notice of adjudication states the dispute as being the amount due in an architect's certificate, and if the redress sought is simply the adjudicator to decide the amount due, the adjudicator will have no power to order payment of that amount, although, doubtless, that would be what the referring party wishes: *F.W. Cook Ltd* v. *Shimizu (UK) Ltd* (2000). The adjudicator can only answer the question posed in the notice of adjudication. Therefore, if the dispute is about payment in the absence of a certificate, and if the redress sought is an order for payment of the amount due, the adjudicator would have no option but to decide that no amount was due and that no order for payment could be made. That is because the employer's duty to pay does not arise until the issue of a certificate stating the amount payable. What the referring party should have done was to ask the adjudicator to decide the amount that should have been certified and to order its payment. The adjudicator is not empowered to answer the question which should have been asked, and an adjudicator who tried to do that would be acting in excess of jurisdiction. The decision would be a nullity.

Paragraph 20, however, allows the adjudicator to take into account any other matters which the parties agree should be within the adjudication's scope. Moreover, the adjudicator is expressly empowered to take into account matters which the adjudicator considers are necessarily connected with the dispute. To take a simple example, it is probably essential for an adjudicator to decide the extent of extension of time allowable, even if not asked, before deciding about the amount of liquidated damages properly recoverable. The express empowerment merely puts into words what would be the legal position in any event: *Karl Construction (Scotland) Ltd* v. *Sweeney Civil Engineering (Scotland) Ltd* (2002); *Sindall Ltd* v. *Solland* (2001).

14.2.3 The Scheme: appointment of the adjudicator

Paragraphs 2 to 6 set out the procedure for selecting an adjudicator. The process is relatively complex (see flowchart in Figure 14.1), which reflects the general nature of the Scheme.

Paragraph 2 contains the framework, and it is made subject to the overriding point that the parties are entitled to agree the name of an adjudicator after the notice of adjudication has been served. If the parties can so agree, they have the best chance of an adjudicator who has the confidence of both parties. Sadly, parties in dispute find it difficult to agree on anything at all. If there is a person named in the contract, that person must first be asked to act as adjudicator. There are difficulties with having a named person: the person may be away or ill or even dead when called upon to act; the person's expertise may be unsuitable for the particular dispute; or pressure of work may force that person to decline.

If there is no named person, or if that person will not or cannot act, the referring party must ask the nominating body indicated in the contract particulars to nominate an adjudicator. There are problems with this approach also. Not all adjudicators are of equal capability. Indeed, some of them are quite poor. Some have a tenuous grasp of the law, and many others believe that the adjudicator's job is to make decisions according to their own gut feeling notions of right and wrong, without reference to law. If the referring party asks for a nomination, both parties are stuck with the result unless they agree to revoke the appointment. However, for the reason already stated, such an agreement is unlikely.

The nominating body is to be stated in the contract particulars. The bodies listed are:

- Royal Institute of British Architects
- The Royal Institution of Chartered Surveyors

- Construction Confederation
- National Specialist Contractors Council
- Chartered Institute of Arbitrators.

Four of the bodies should be deleted. If there is no adjudicator named, and no body is selected, the referring party may choose any one of the bodies to make the nomination. If there is no list of appointing bodies, perhaps because all of them have inadvertently been deleted, the referring party is free to choose any nominating body to make the appointment. A nominating body is fairly broadly defined in paragraph 2(3) as a body which holds itself out publicly as a body which will select an adjudicator on request. The body may not be what is referred to as a 'natural person', i.e. a human being, nor one of the parties. Paragraph 6 states that if an adjudicator is named in the contract, but for some reason cannot act or does not respond, the referring party has three options. The first is to ask any other person specified in the contract to act. The second is to ask the adjudicator-nominating body in the contract to nominate, and the third is to ask any other nominating body to nominate. It will readily be seen that this procedure is simply a clarification of existing options.

The nominated adjudicator has two days in which to accept, from receiving the request. The adjudicator must be a single person and not a body corporate. Therefore a firm of quantity surveyors cannot be nominated, although one of the directors or partners can be nominated. Paragraph 5 provides that the nominating body has five days from receipt of the request to communicate the nomination to the referring party. Invariably, a nominating body will also notify the respondent, but surprisingly the Scheme does not expressly require such notification.

In the event that the nominating body fails to nominate within five days, either the parties may agree on the name of an adjudicator or the referring party may request another nominating body to nominate. In either case the adjudicator has two days to respond as before.

Paragraph 10 makes clear that the objection of either party to the adjudicator will not invalidate the appointment, nor any decision reached by the adjudicator. It is useful to have this spelled out. There is a misconception among the uninitiated that a party has only to register an objection to the adjudicator in order to bring the process to an end or at least suspend it. Nothing could be further from the truth. It is entirely a matter for each party the extent to which it wishes to participate in the adjudication. If a party objects, it should make quite clear that any participation is without prejudice to that position and to the party's right to refer the objection to the courts in

Figure 14.1
Selection of the adjudicator

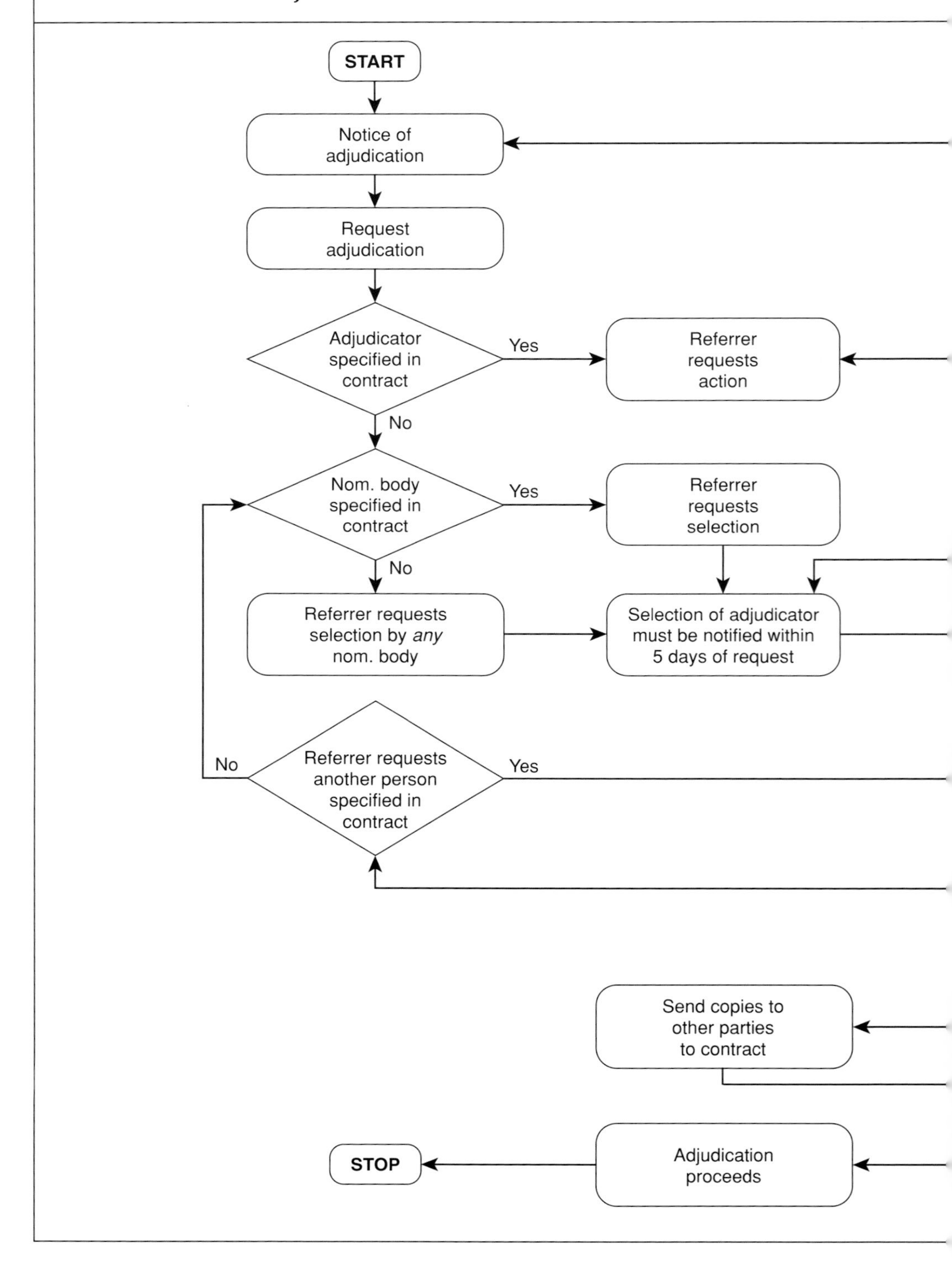

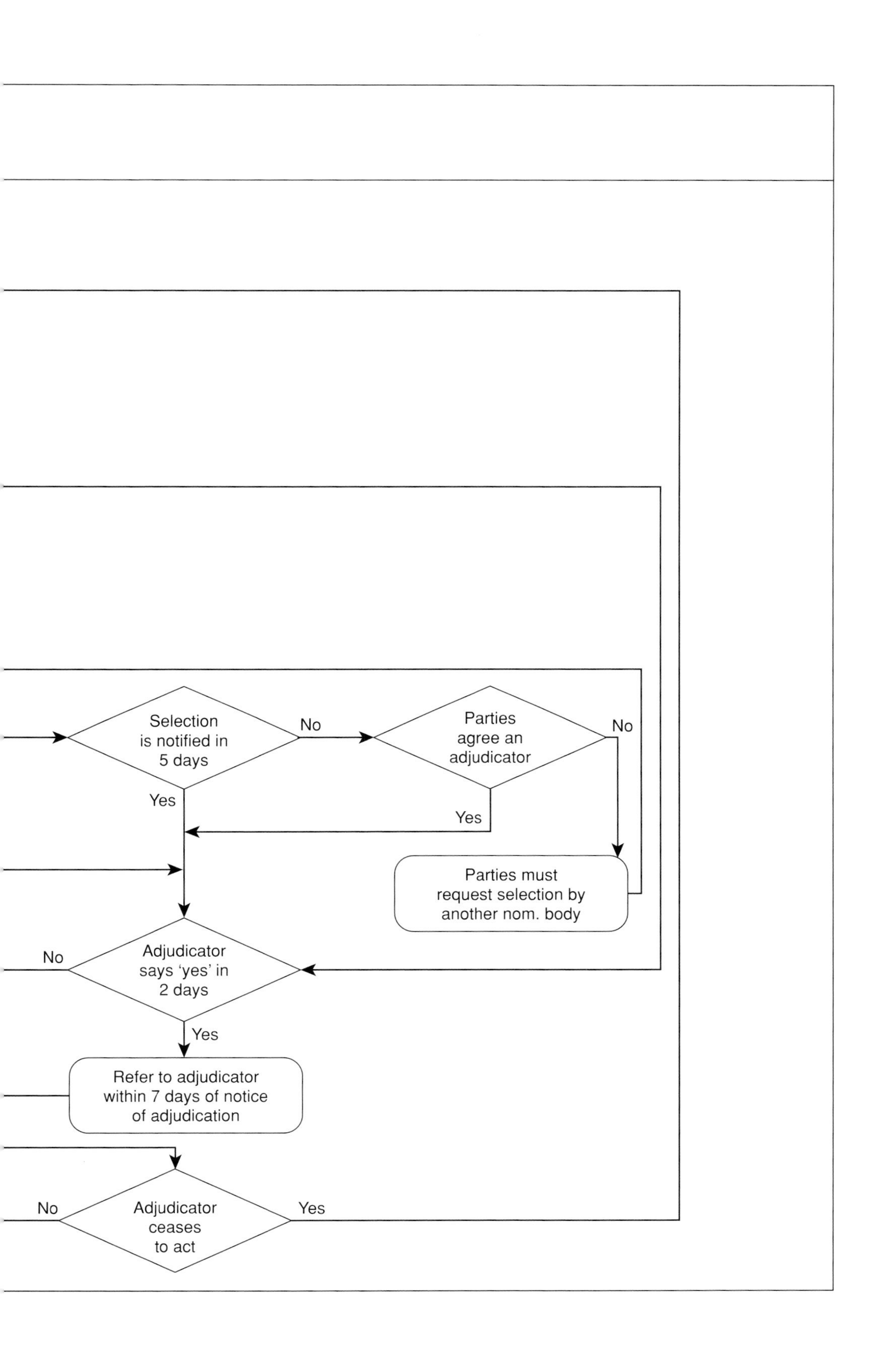

Selection is notified in 5 days
No
Parties agree an adjudicator
No
Yes
Yes
Parties must request selection by another nom. body
No
Adjudicator says 'yes' in 2 days
Yes
Refer to adjudicator within 7 days of notice of adjudication
No
Adjudicator ceases to act
Yes

due course. It may be catastrophic for a party knowing of grounds for objection to continue the adjudication without further comment. In such cases the party may well be deemed to have accepted the adjudicator: *R. Durtnell & Sons Ltd* v. *Kaduna Ltd* (2003).

The Scheme makes elaborate provision in paragraphs 9 and 11 if the adjudicator resigns or the parties revoke the appointment. The provisions are sensible. The adjudicator may resign at any time on giving notice in writing to the parties. Note that the notice need not be reasonable, and therefore immediate resignation is possible. The referring party may serve a new notice of adjudication and seek the appointment of a new adjudicator, as noted above. If the new adjudicator so requests, and if reasonably practicable, the parties must make all the documents available which have been previously submitted.

An adjudicator who finds that the dispute is essentially the same as a dispute which has already been the subject of an adjudication decision must resign. The adjudicator is entitled to determine a reasonable amount due by way of fees and expenses and how it is to be apportioned. The parties are jointly and severally liable for any outstanding sum. Although, oddly, not given as a reason for resignation, the significant variation of a dispute from what was referred in the referral notice, so that the adjudicator is not competent to decide it, is a trigger for entitlement to payment. At first sight this criterion is not easy to understand. The only sensible interpretation is that it appears to refer to a significant difference between the dispute as set out in the notice of adjudication and what the referring party then includes in the referral notice.

Revocation of the appointment by the parties under paragraph 11 does not seem to be a common occurrence. When it occurs, the adjudicator is entitled to determine a reasonable amount of fees and expenses, and the apportionment. The parties, as before, are jointly and severally liable for any balance. Parties will find it difficult to challenge the amount of fees determined by an adjudicator unless the adjudicator can be shown to have acted in bad faith. It is not sufficient to show that a court would have arrived at a different sum: *Stubbs Rich Architects* v. *WH Tolley & Son* (2001).

14.2.4 The Scheme: procedure

The procedure is indicated by the flowchart in Figure 14.2. Paragraph 3 stipulates that a request for the appointment of an adjudicator must include a copy of the notice of adjudication. As the court stated in the *Ken Griffin* case, this is to assist those making the nomination and the prospective adjudicator so that a suitable person is nominated. There is little time available because, in compliance with the Act, paragraph 7 stipulates that the referring party must submit the dispute in writing to the adjudicator, with copies to each party to the dispute, no later than seven days after the notice of adjudication. This submission is known as the 'referral notice'. Looking at the procedure for appointment of the adjudicator, it is clear that the timetable is tight.

The referral notice, which effectively is the referring party's claim, must be accompanied by relevant parts of the contract and whatever other evidence the referring party relies upon in support of the claim.

The Scheme does not indicate that the respondent may reply to the referral notice, but in order to comply with the rules of natural justice the adjudicator is obliged to allow a reasonable period for the reply. The provisions in IFC 98 used to allow seven days for the reply. Respondents always believe this to be totally inadequate to reply to what may be a referral notice and evidence amounting to several lever arch files. It is a matter for the adjudicator to decide, but in view of the restricted overall period for the decision it seems that 14 days is the very most that any respondent can expect. Usually, the adjudicator will allow less than this. Paragraph 19 provides that the adjudicator must reach a decision 28 days after the date of the referral notice. Note that this is from the date of the notice, not from the date the notice is received by the adjudicator. The period may be extended by 14 days if the referring party consents or, if both parties agree, for any longer period.

If the adjudicator does not comply with this timetable in reaching the decision, either party may serve a new notice of adjudication and request a new adjudicator to act. The new adjudicator can request copies of all documents given to the former adjudicator. Paragraph 19(3) requires the adjudicator to deliver a copy of the decision to the parties as soon as possible after the decision has been reached.

Questions have arisen about the position if the adjudicator does not reach a decision in time or if the decision reached in time is not delivered as soon as possible. There are two conflicting Scottish decisions and two English decisions dealing with these questions. *Ritchie Brothers PWC Ltd* v. *David Philp (Commercials) Ltd* (2004) decided that the adjudicator's jurisdiction to make a decision ceased on the expiry of the time limit if not already extended. *St Andrews Bay Development Ltd* v. *HBG Management Ltd* (2003) decided that, if the adjudicator reached a decision within the relevant timescale,

Figure 14.2
Procedure under Scheme and relation to arbitration

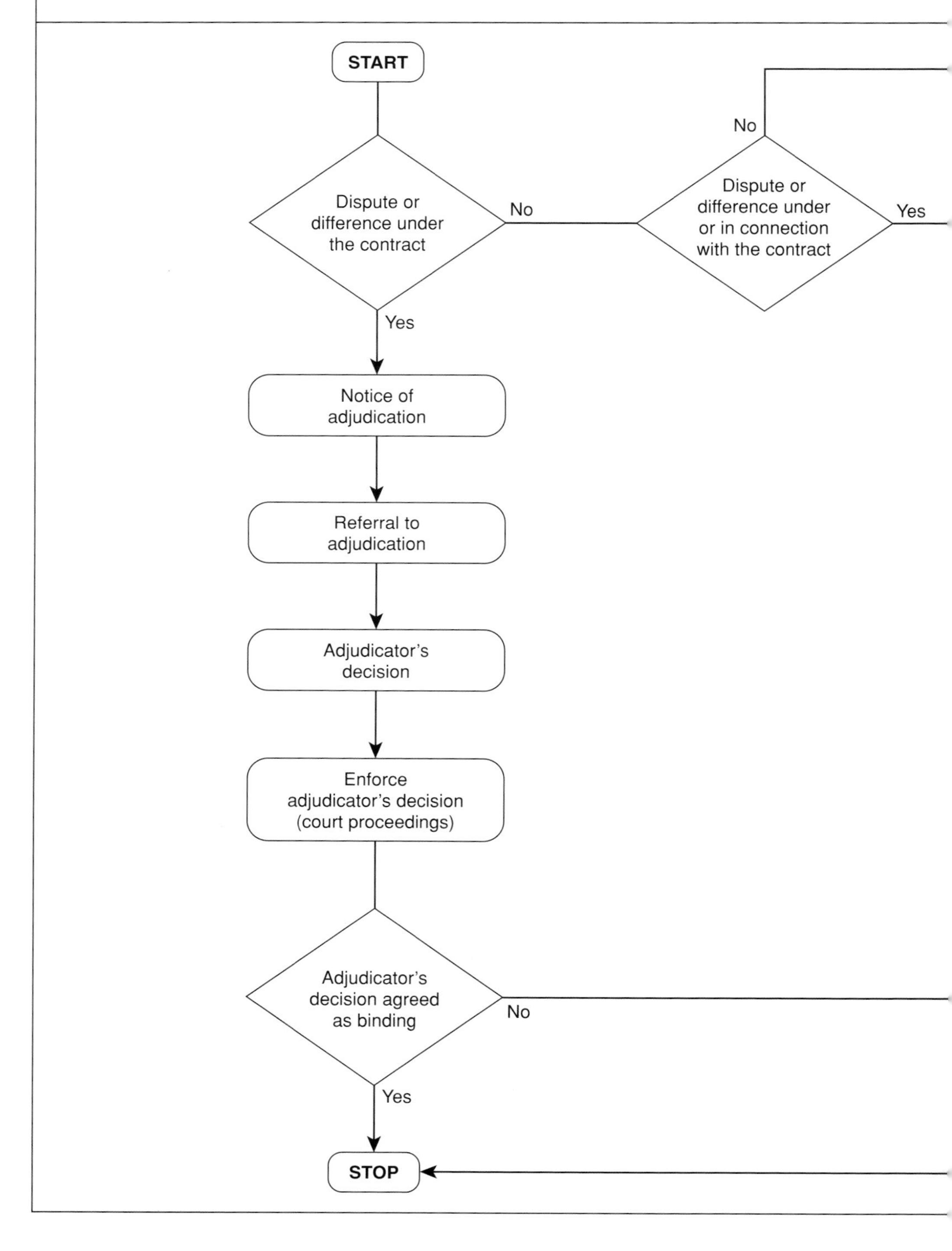

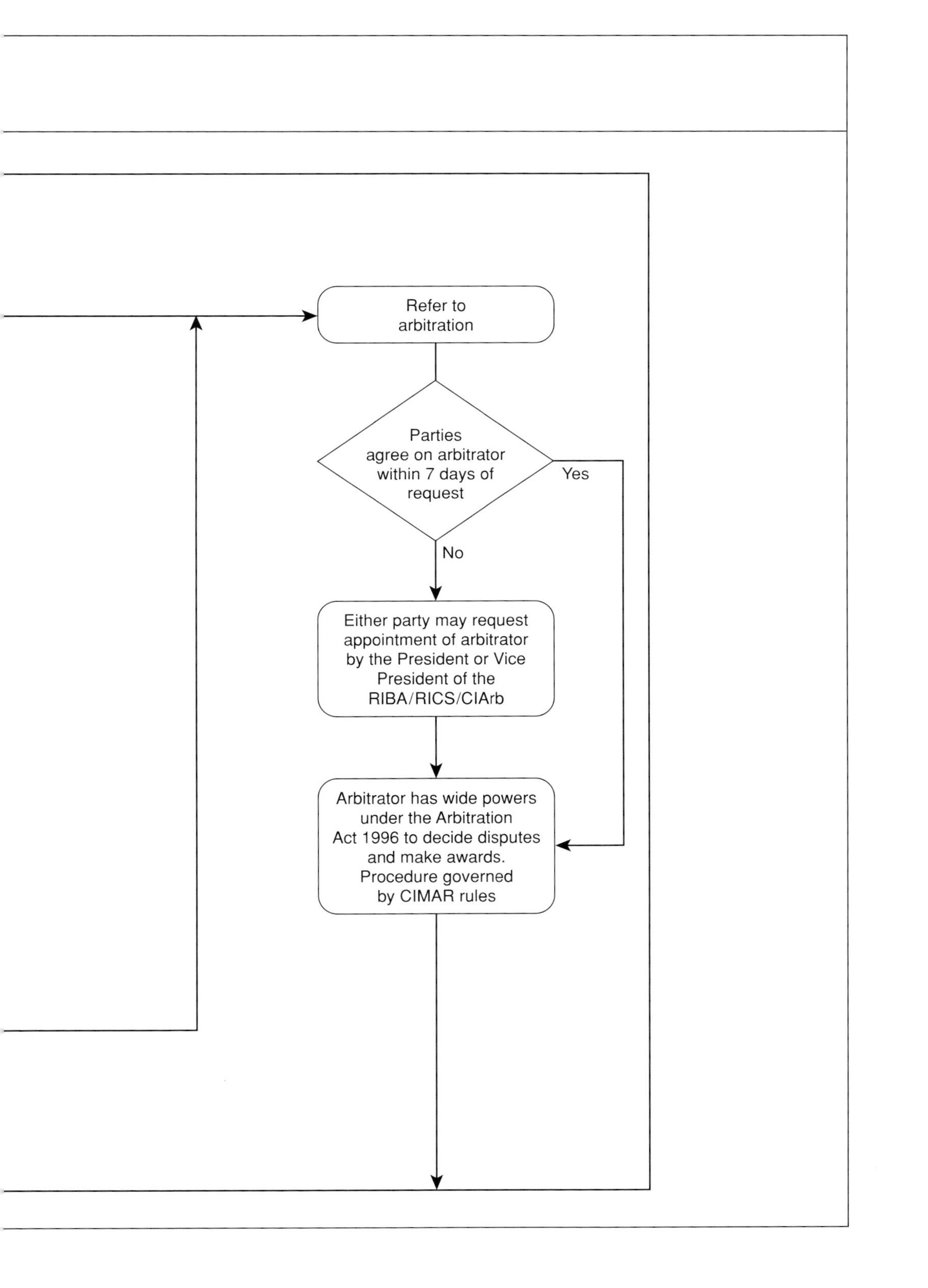
Refer to arbitration
Parties agree on arbitrator within 7 days of request
Yes
No
Either party may request appointment of arbitrator by the President or Vice President of the RIBA/RICS/CIArb
Arbitrator has wide powers under the Arbitration Act 1996 to decide disputes and make awards. Procedure governed by CIMAR rules

a two-day delay in delivering the decision to the parties was not sufficient to render the decision a nullity. *Simons Construction Ltd* v. *Aardvark Developments Ltd* (2004) decided that the decision of an adjudicator is binding, even if given after the expiry of the relevant timescale provided that neither party has served a new notice of adjudication before the decision has been reached. *Barnes & Elliott Ltd* v. *Taylor Woodrow Holdings Ltd* (2004) decided that if the adjudicator reached the decision within the relevant timescale, a short delay in communicating it to the parties was within the tolerance and commercial practice that should be afforded to the Act and the contract. The Scottish decisions are not, of course, binding in English courts, but they may be persuasive if there is no English decision on the point. As there are English decisions, the position appears to be as set out in those decisions: a late or late-communicated decision is valid provided neither party has taken steps to bring the adjudication to an end such as serving a fresh notice of adjudication after the expiry of the relevant period.

If one of the parties fails to comply with the adjudicator's decision, the other may seek enforcement of the decision through the courts. The courts will normally enforce the decision unless there is a jurisdictional or procedural problem. In enforcement proceedings the court is not being asked to comment on the adjudicator's decision or reasoning, although a court will quite often do so, thus obscuring the *ratio* of the judgment. Where a court is asked to enforce an adjudicator's decision, the important part of the judgment is simply the reasons why the judge decided to enforce or not. Any comments the judge may make on the adjudicator's decision itself will be *obiter*: at best persuasive, but certainly not of binding force.

14.2.5 The Scheme: adjudicator's powers and duties

Paragraph 8 permits the adjudicator to adjudicate at the same time on more than one dispute under the same contract, provided that all parties consent. The adjudicator may deal with related disputes on several contracts even if not all the parties are parties to all the disputes, provided they all consent. Moreover, the parties may agree to extend the period for decision on all or some of the disputes. It is clear that multiple dispute procedures bring their own complications, for which the Scheme, wisely, does not try to legislate. For example, it is not clear whether multiple disputes, certainly under different contracts, must be adjudicated on under one big adjudication. The wording of paragraph 8 seems to leave the position open.

Where there are different contracts, and the parties vary from one contract to another, it will be a matter of discussion and agreement whether the adjudicator should conduct separate adjudications, albeit at the same time. It is quite conceivable that parties to different contracts may not wish others to know some of the details of the dispute. Of course, if the disputes were being dealt with through the courts, not only the parties themselves but any member of the public would be able to sit in and listen to those details.

Another problem with multiple adjudications is that an adjudicator may have been appointed for one or more of the adjudications. If so, and that adjudicator ceases to act under this paragraph, the adjudicator concerned is entitled to render a fee account to the relevant parties, who will be jointly and severally liable for its discharge if no apportionment is made or if one party fails to pay some or all of its share.

The adjudicator's duties are to act impartially in accordance with the relevant contract terms, to reach a decision 'in accordance with the applicable law in relation to the contract', and to avoid unnecessary expense. Sadly, some adjudicators seem to be unaware of their obligations to apply the law to their decisions, and decisions are made on the basis of the adjudicators' idea of fairness, moral rights or justice. The author has had the misfortune to see some appalling adjudicator's decisions. In one instance the adjudicator decided that the executed contract was unsuitable for the Works, and drafted the decision as though a different contract had been used. Needless to say, the decision was declared a nullity by the court. On another occasion an adjudicator declared himself a 'watchdog for fairness'. That is a misguided view of the adjudicator's role, which has been stated to be 'primarily to decide the facts and apply the law (in the case of an adjudicator, the law of the contract)': *Glencot Development and Design Company Ltd* v. *Ben Barrett and Son (Contractors) Ltd* (2001). Fortunately, there are also some very good adjudicators with a clear understanding of their roles.

The adjudicator is given some very broad and some very precise powers:

- To take the initiative in ascertaining the facts and the law
- To decide the procedure in the adjudication
- To request any party to supply documents and statements
- To decide the language of the adjudication and order translations
- To meet and question the parties
- To make site visits, subject to any third party consents
- To carry out any tests, subject to any third party consents

- To obtain any representations or submissions
- To appoint experts or legal advisers, subject to giving prior notice.
- To decide the timetable, deadlines and limits to length of documents or oral statements
- To issue directions about the conduct of the adjudication.

Paragraph 14 requires parties to comply with the adjudicator's directions. If a party does not comply, the adjudicator has significant powers (paragraph 15):

- To continue the adjudication notwithstanding the failure
- To draw whatever inferences the adjudicator believes are justified in the circumstances
- To make a decision on the basis of the information provided and to attach whatever weight to evidence submitted late the adjudicator thinks fit.

Paragraph 16 deals with representation. A party may have assistance or representation as deemed appropriate, with the proviso that oral evidence or representation may not be given by more than one person unless the adjudicator decides otherwise.

Very occasionally, a contract may provide that a decision or certificate is final and conclusive. Unless that is the case, the adjudicator is given power to open up, revise and review any decision or certificate given by a person named in the contract. It is worth noting that to be exempt from revision by the adjudicator, the decision or certificate must be stated to be both final and conclusive. A contract that simply states that a certificate is conclusive is open to review. On that basis, the final certificate is exempt because it is called 'final' and stated to be conclusive. Obviously it is reviewable if the reference is made before the expiry of 28 days from the date of issue in accordance with clause 1.10.2. The adjudicator is also given power to order any party to the dispute to make a payment, its due date and the final date for payment and to decide the rates of interest, the periods for which it must be paid and whether it must be simple or compound interest. In deciding what, if any, interest must be paid, the adjudicator must have regard to any relevant contractual term. To 'have regard' to a contractual term is a rather loose phrase, which probably means little more than to give attention to it. It falls short of the need to actually comply with it: *R* v. *Greater Birmingham Appeal Tribunal ex parte Simper* (1973).

The adjudicator must consider relevant information submitted by the parties, and if the adjudicator believes that other information or case law should be taken into account, it must be provided to the parties and they must have the opportunity to comment: *Balfour Beatty Construction Ltd* v. *London Borough of Lambeth* (2002). Paragraph 18 puts a prohibition on the disclosure of information, noted by the supplier as confidential, to third parties by any party to an adjudication or the adjudicator unless the disclosure is necessary for the adjudication.

14.2.6 The Scheme: the adjudicator's decision

Paragraph 23 empowers the adjudicator to order the parties to comply peremptorily with the whole or any part of the decision. In the absence of any directions about the time to comply, paragraph 26 makes clear that compliance must be immediate on delivery of the decision to the parties.

The Scheme repeats the requirement of the Act that the decision will be binding and must be complied with until the dispute is finally determined by arbitration, legal proceedings or agreement.

In contrast to the position under IFC 98, paragraph 22 provides that, if either party so requests, the adjudicator must give reasons for the decision. Some adjudicators purport not to give reasons, but simply indications or limited reasons. The court in *Joinery Plus Ltd (in administration)* v. *Laing Ltd* (2003) had some useful things to say about reasons:

> 'The statement by the adjudicator that he was only giving reasons for a limited purpose or was only giving limited reasons has little if any practical effect. If an adjudicator gives any reasons, they are to be read with the decision and may be used as a means of construing and understanding the decision and the reasons for that decision. There is no halfway house between giving reasons and publishing a silent or non-speaking decision without any reasons. There is no way in which reasons may be given for a limited purpose and which are only capable of being used for that purpose.'

The court went on to say that comments about the decision given by the adjudicator after delivering the decision were irrelevant except to the extent that the adjudicator was entitled to correct basic mistakes in the decision, if invited to do so: *Bloor Construction (UK) Ltd* v. *Bowmer & Kirkland (London) Ltd* (2000).

The adjudicator is entitled to reasonable fees, which the adjudicator may determine. The parties are jointly and severally liable for payment if the adjudicator makes no apportionment, or if there is an outstanding balance.

Paragraph 26 states that the adjudicator will not be liable for anything done or omitted in carrying out the functions of an adjudicator unless the act or omission is in bad faith. Similar protection is also given to any employee or agent of the adjudicator. It is perhaps worth noting that, as an incorporated term of the contract, this paragraph is not binding on persons who are not parties to the contract.

14.2.7 The Scheme: costs

Nothing in the Scheme allows the adjudicator to award the parties costs. This is in harmony with the philosophy of the Act, which does not encourage the parties to incur large amounts of costs in pursuing claims. In arbitration and litigation, by contrast, where costs are normally awarded against the losing party, the dispute can deteriorate into a fight about costs rather than about the point at issue. That has much to do with the huge costs which can be incurred by each side.

Despite this, whether or not, in a particular instance, an adjudicator can award costs has caused problems. *John Cothliff Ltd* v. *Allen Build (North West) Ltd* (1999), which decided that the Scheme gave the adjudicator power to award costs, was considered in *Northern Developments (Cumbria) Ltd* v. *J. & J. Nichol* (2000). There the court did not agree with the earlier case, and in a concise judgment held that there was no provision in the Scheme that gave the adjudicator such power. However, the adjudicator could be given power to award costs, either expressly by the parties or by implied agreement. In that case, both parties had professional representation. Both parties asked the adjudicator to award costs, and neither party made any submissions that the adjudicator had no power to award costs. As the judge said: 'It would have been open to either party to say to the Adjudicator, I have only asked for costs in case you decide that you have jurisdiction to award them but I submit that you have no jurisdiction to make such an award.'

14.3 *Arbitration*

14.3.1 General

Until the House of Lords' decision in *Beaufort Developments (NI) Ltd* v. *Gilbert-Ash NI Ltd* (1998), it was held that a court did not have the same power as the arbitrator, conferred by the contract, to open up and revise certificates and decisions: *Northern Regional Health Authority* v. *Derek Crouch Construction Co Ltd* (1983). Undoubtedly, arbitration was chosen over legal proceedings in many contracts on that basis. That was because both employers and contractors were afraid that, if matters were left to litigation, disputes about certification (which the employer as well as the contractor might wish to challenge) could not properly be resolved.

That approach was changed by the *Beaufort* case, which held that a court has the same powers as an arbitrator to open up, review and revise certificates, opinions and decisions of the architect. Indeed, the court has it as a right, whereas the power must be conferred upon the arbitrator by the parties. It has been remarked at the beginning of this chapter that legal proceedings are now the default procedure in IC and ICD, rather than arbitration.

IC arbitration procedures are very brief. Arbitration can take place on any matter at any time. Arbitrators appointed under a JCT arbitration agreement are given extremely wide express powers. Their jurisdiction is to decide any dispute or difference of any kind whatsoever arising under the contract or connected with it (article 8). The scope could scarcely be broader (*Ashville Investments Ltd* v. *Elmer Contractors Ltd* (1987)), and by clause 9.5 the arbitrator's powers extend to:

- Rectification of the contract to reflect the true agreement between the parties
- Directing the taking of measurements or the undertaking of such valuations as the arbitrator thinks desirable to determine the respective rights
- Ascertaining and making an award of any sum which should have been included in a certificate
- Opening up, reviewing and revising any certificate, opinion, decision, requirement or notice issued, given or made, and determining all matters in dispute as if no such certificate, opinion, decision, requirement or notice had been given.

The JCT 2005 edition of the Construction Industry Model Arbitration Rules (CIMAR) current at the contractual base date are to govern the proceedings (clause 9.3). The provisions of the Arbitration Act 1996 are expressly stated by clause 9.8 to apply to any arbitration under this agreement. That is to be the case no matter where the arbitration is conducted. Therefore, even if the project and the arbitration take place in a foreign jurisdiction, the UK Act will apply provided that the parties contracted on IC and clause 1.12 referring to the law of England is not amended.

The following matters are specifically excluded from arbitration:

- Disputes about value added tax
- Disputes under the Construction Industry Scheme, where legislation provides some other method of resolving the dispute
- The enforcement of any decision of an adjudicator.

The employer and contractor agree, by clause 9.7 in accordance with sections 45(2)(a) and 69(2)(a) of the Act, that either party may by proper notice to the other and to the arbitrator apply to the courts to determine any question of law arising in the course of the reference, and appeal to the courts on any question of law arising out of an award. When the clause was originally introduced it was viewed with some doubt, on the basis that the courts might not accept it as satisfying the requirements prior to such an appeal. However, clauses like this have been held to be effective: *Vascroft (Contractors) Ltd* v. *Seeboard plc* (1996).

Arbitration, like litigation, is almost always costly in terms of both money and time. No matter how powerful and convincing the case may be, there is no guarantee of success. Even the successful party will often look back at the cost, the time spent and the mental stress involved and conclude that it was not worth the effort.

There are always some people who will threaten arbitration over trivial matters in an attempt to gain an advantage. Indeed, it is a recognised, although possibly ineffective, form of negotiation to suddenly abandon talks and serve an arbitration notice. Unfortunately, even with the recent review of dispute resolution procedures and introduction of the adjudication process, that approach will not disappear. It will not always be possible to avoid arbitration, and therefore employers and contractors must ensure that they properly appreciate how the process operates. Only then can they recognise the possible consequences of embarking on formal arbitration proceedings.

A common misconception is that the arbitration process is an informal get-together to enable the parties and the arbitrator to have a chat about the dispute before the arbitrator decides, in a consensus kind of a way, who should be successful. That is much more a description of a conciliation meeting. In fact, the majority of arbitrations are conducted quite formally, like private legal proceedings – which is what they are. The arbitration begins by inviting the parties to the 'preliminary meeting', but that does not mean a friendly discussion. It is a formal meeting to establish all the important criteria which need to be decided before the arbitration can proceed. The arbitrator normally works from an agenda. Sometimes parties attempt to gain an advantage by springing a surprise request on the arbitrator at that meeting. Experienced arbitrators have no difficulty in dealing with such requests, but there is a limit to the degree to which the arbitrator can ensure that one party is not disadvantaged by such tactics. A party should not go to a preliminary meeting without taking a fully briefed legal adviser experienced in arbitration.

The employer and contractor are free to agree who should be appointed as, or should appoint, the arbitrator, and they have freedom to agree important matters such as the form and timetable of the proceedings. This raises the possibility of a quicker procedure than would otherwise be the case in litigation, and even matters such as the venue for any future hearing might be arranged to suit the convenience of the parties and their witnesses.

If oral evidence and cross-examination is to be carried out, it is usually done at a hearing. Hearings, which are the private equivalent of a trial, are conducted in private, not in an open court. Parties are free to choose whether to represent themselves or whether to be represented and by whom. They need not, in the traditional courtroom way, be represented by solicitor and counsel. It is not usually advisable for the parties to represent themselves, because difficult legal points can arise in apparently the simplest of arbitrations. Some arbitrations are won by clever tactics. Therefore experienced help is essential.

14.3.2 Procedure

Arbitrations begun under IC or ICD and subject to the law of England must be conducted subject to and in accordance with the JCT 2005 edition of CIMAR, current at the base date of the contract. If any amendments have been issued by JCT since that date, the parties may jointly agree to give written notice to the arbitrator to conduct the reference according to the amended rules (clause 9.3). CIMAR is a comprehensive body of rules, generally of admirable clarity. It consists of a set of rules with two appendices: the first defines terms and the second helpfully reproduces sections of the Arbitration Act 1996 which are relevant, but not already included in the rules. These are followed by the JCT Supplementary and Advisory Procedures, the first part of which is mandatory and must be read with the rules. The greater part is advisory only, but appears to be well worth adopting. At the back of the document is a set of notes prepared by the Society of Construction Arbitrators dated 1 February 1998 and updated January 2002. The whole document, at the time of writing, is available on *www.jctcontracts.com*. As might be expected, JCT/CIMAR is very detailed and repays careful study. Among other things, it offers the parties a choice of three broad categories of procedure by which the proceedings will be conducted, as follows.

Short hearing procedure (rule 7)

This procedure is not very common. It limits the time available to the parties within which to orally address the matters in dispute before the arbitrator. Although the time can be extended by mutual consent, in the absence of that agreement no more than one day will be allowed, during which both parties must have a reasonable opportunity to be heard. Before the hearing, either by simultaneous exchange or by consecutive submissions, each party will provide to the arbitrator and to each other a written statement of their claim, defence and counterclaim (if any). Each statement must be accompanied by all relevant documents and any witness statements on which it is proposed to rely. The JCT procedures usefully insert some timescales for certain of the steps.

If it is appropriate to do so, either before or after the short hearing, the arbitrator may inspect the subject matter of the dispute if desired. This procedure is particularly suited to issues which can be decided fairly easily by such an inspection of work, materials, plant and/or equipment or the like. The arbitrator must decide the issues and make an award within a month after concluding hearing the parties.

It is possible to present expert evidence. However, it is costly and often unnecessary. That is particularly the case if the arbitrator has been chosen specifically on the grounds of specialist knowledge and expertise. Parties can sometimes agree to allow the arbitrator to use that specialist expertise when reaching the decision, and so the use of independent expert evidence under the short hearing procedure is all but actively discouraged under rule 7.5, which precludes any party calling such expert evidence from recovering the costs of doing so, except where the arbitrator determines that such evidence was 'necessary for coming to his decision'.

This procedure with a hearing is ideally suited to many common disputes which are relatively simple, and provides for a quick award with minimum delay and associated cost.

Documents only procedure (rule 8)

This will rarely be a viable option. It is not viable unless all the evidence is contained in the form of documents. Nevertheless, if the criteria are satisfied, it can offer real economies of time and cost. It is best suited to disputes which are capable of being dealt with in the absence of oral evidence, and where the sums in issue are relatively modest and do not warrant the time and associated additional expense of a hearing. Parties, in accordance with a timetable devised by the arbitrator, will serve on each other and on the arbitrator a written statement of case which, as a minimum, will include:

- An account of the relevant facts and opinions upon which reliance is placed
- A statement of the precise relief or remedy sought.

If either party is relying on evidence of witnesses of fact, the relevant witness statements (called 'proofs'), signed by the witnesses concerned, will be included with the statement of case. If the opinions of an expert or experts are required, they must similarly be given in writing and signed. There is a right of reply, and if there is a counterclaim, the other party may reply to it.

Despite the title of this procedure, the arbitrator may set aside up to a day during which to question the parties and/or their witnesses if it is considered desirable. The arbitrator must make a decision within a month or so of final exchanges and questioning, but there is provision for the arbitrator to notify the parties that more time for the decision will be required. The JCT procedures again set out a useful timetable.

Full procedure (rule 9)

If neither of the other options is considered satisfactory, CIMAR makes provision for the parties to conduct their respective cases in a manner similar to conventional High Court proceedings, offering the opportunity to hear and cross-examine factual and expert witnesses.

This is the most complex procedure, and the JCT procedure which sets out a detailed timetable for various activities within the procedure is of real assistance to the parties and to the arbitrator. It is intended that the rules will accommodate the whole range of disputes which might arise. Therefore they offer a sensible framework for conducting the proceedings. They may be modified as appropriate so that they can be used effectively and efficiently for the particular dispute under consideration.

The unamended rules lay down that parties will exchange formal statements. In difficult or complex cases the statements will comprise the claim, defence and counterclaim (if any), reply to defence, defence to counterclaim and reply to defence to counterclaim. Each submission must be sufficiently detailed to enable the other party to answer each allegation made. As a minimum, the statement must set out the facts and matters of opinion which are to be established by evidence. It may include statements concerning any relevant points of law, evidence, or reference to the evidence that it is proposed will be presented, if this will assist in defining the issues and a clear statement of the relief or remedies sought.

The arbitrator should give detailed directions concerning everything necessary for the proper conduct of the arbitration. Often, those directions will include orders regarding the time within which either party may request further and better details of the other party's case and the reply to any such request. Directions may also be given requiring the disclosure of any documents or other relevant material which is or has been in each party's possession. Probably, the parties will be required to exchange written statements setting out any evidence that may be relied upon from witnesses of fact in advance of the hearing. There will also be directions given regarding any expert witnesses, the length of the hearing or hearings, and the time available for each party to present its case.

14.3.3 The appointment of an arbitrator

It is at the option of either party to begin arbitration proceedings. As a first step, one party must write to the other requesting it to concur in the appointment of an arbitrator (clause 9.4.1). Whoever does so, proceedings are formally commenced when the written notice is served. Rule 2.1 of CIMAR sets out the procedure, stating that the notice must identify the dispute and require agreement to the appointment of an arbitrator. It is good practice for the party seeking arbitration to insert the names of three prospective arbitrators. This saves time, and often both parties can agree on one of the names. Figure 14.3 is a suitable letter. If the respondent maintains that none of the names is acceptable, it is usual for that party to volunteer a new set of names. The arbitrator must have no relationship to either of the parties, nor should the arbitrator have connections with any matter associated with the dispute.

Figure 14.3
Letter requesting concurrence in appointment of arbitrator

SPECIAL DELIVERY

Dear

We hereby give you notice that we require the undermentioned dispute or difference between us to be referred to arbitration in accordance with article 8 and clause 9.3 of the contract dated [*insert date*]. Please treat this as a request to concur in the appointment of an arbitrator under clause 9.4.1.

The dispute or difference is: [*specify*].

I propose the following three persons for your consideration and require your concurrence in the appointment within 14 days of the date of receipt of this letter, failing which we shall apply to the President of the Royal Institute of British Architects/The Royal Institution of Chartered Surveyors/Chartered Institute of Arbitrators [*delete as appropriate*] for the appointment of an arbitrator.

The names and addresses we propose are: [*insert names and addresses*].

Yours faithfully

It is important for the parties to make a sincere effort to agree on a suitable candidate rather than having one appointed whose skills and experience may be entirely unknown. Depending on the nature and amount of money at stake in the dispute, the parties may be unwilling to agree anything. In that case, clause 9.4.1 of the contract and rule 2.3 of CIMAR provide that, if the parties cannot agree upon a suitable appointment within 14 days of a notice to concur or any agreed extension to that period, either party can apply to a third party to appoint an arbitrator. There is a list of appointers in the contract particulars against clause 9.4.1. All but one should be deleted, leaving the agreed appointer as either the President or a Vice-President of the Royal Institute of British Architects, The Royal Institution of Chartered Surveyors, or the Chartered Institute of Arbitrators. If no single body has been chosen, the default provision is the President or a Vice-President of the Royal Institute of British Architects. Of course, it is always open to the parties to the contract to insert the name of a different appointer of their choice at the time the contract is executed. Figure 14.4 is the sort of letter that a contracting party might send to an appointing body requesting appointment. However, it will be necessary to complete special forms and to pay the relevant fee. Although the system of appointing an arbitrator varies, the aim is the same. The object is to appoint a person of integrity who is independent, having no existing relationships with either party or their professional advisers, and who is impartial. It should go without saying that the arbitrator should have the necessary and appropriate technical and legal expertise. Claimants who have a dispute to refer and respondents receiving a notice to concur should waste no time in taking proper expert advice on how best to proceed.

If the arbitrator's appointment is made by agreement, it will not take effect until the appointed person has confirmed willingness to act, irrespective of whether terms have been agreed. If the appointment is the result of an application to the appointing body, it becomes effective, whether or not terms have been agreed, when the appointment is made by the relevant body (CIMAR rule 2.5). There is no fixed scale of charges for arbitrators' services, and fees ought to depend on their experience, their expertise, and often on the complexity of the dispute. Arbitrators commonly charge between £1000 and £2000 a day. They usually require an initial deposit from the parties, and, if there is to be a hearing, there will be a cancellation charge graded in accordance with the proximity of the cancellation to the start of the hearing. The argument in support of this is that the arbitrator will have put one or two weeks on one side for

Figure 14.4
Letter to relevant appointing body for arbitrators

SPECIAL DELIVERY

Dear

We are the employer/contractor [*delete as appropriate*] under an IC/ICD [*delete as appropriate*] form of contract which makes provision for your President to appoint an arbitrator in default of agreement.

We should be pleased to receive the appropriate application form and supporting information, with a note of the current fee payable on application.

Yours faithfully

the hearing, during which time no other work has been booked. A cancellation means that it is difficult for the arbitrator to secure work at short notice to fill the void. In cases where the cancellation fee is substantial, owing to proximity to the hearing date, it might be sensible to ask the arbitrator to account to the parties for activities during the hearing period.

After appointment, the arbitrator will consider which of the procedures summarised above appears to be most appropriate as a forum for the parties to put their cases. The arbitrator must choose the format that will best avoid undue cost and delay, and this is often a most difficult balancing act. Therefore parties must, within 14 days after acceptance of the appointment is notified to the parties, provide the arbitrator with an outline of their disputes and of the sums in issue, along with an indication of which procedure they consider best suited to them. After due consideration of all parties' views, and unless a meeting is considered unnecessary, the arbitrator must, within 21 days of the date of acceptance, arrange a meeting (the preliminary meeting), which the parties or their representatives will attend to agree (if possible) or receive the arbitrator's decision upon everything necessary to enable the arbitration to proceed. It is obviously preferable for the parties to agree which procedure is to apply. If they cannot agree, the documents only procedure will apply unless the arbitrator, after having considered all representations, decides that the full procedure will apply.

The parties are always free to conduct their own cases, but if disputes have reached the stage of formal proceedings it is usually better to engage experienced professionals to act for them.

14.3.4 Counterclaims

Under CIMAR rule 2.1 and s. 14(4) of the Arbitration Act 1996, arbitral proceedings are begun in respect of a dispute when one party serves on the other a written notice of arbitration 'identifying the dispute' and requiring agreement to the appointment of an arbitrator. That is important for two particular reasons. The first is that it is relevant in terms of the Limitation Act 1980. If the notice is served before the expiry of the period, it prevents the respondent using the limitation defence. The second reason is that notice of arbitration often results in a counterclaim from the respondent. It seems doubtful, strictly, whether any counterclaim might be brought within the jurisdiction of the arbitration which has already commenced without a formal process having been executed.

It is common for respondents simply to raise their counterclaims formally at or about the time of serving their defences to the primary claim. That may not be possible in the face of an attack by a claimant wishing to frustrate the respondents' attempts to automatically bring that counterclaim into the proceedings. Note that CIMAR rule 3.2 allows any party to an arbitration to give notice in respect of any other dispute. Provided it is done before the arbitrator is appointed, the disputes are to be consolidated. Rule 3.3 allows either party to serve notice of any other dispute after the arbitrator has been appointed, but consolidation is not then automatic. Rule 3.6 of CIMAR makes clear that arbitral proceedings in respect of any other dispute are begun 'when the notice of arbitration for that other dispute is served'. Although a claimant's insistence that the respondent serve a fresh notice to cover a counterclaim may be only a temporary inconvenience to the respondent, there are serious practical issues.

If there is a doubt concerning whether a counterclaim has properly been brought within the original arbitration, it may affect the existence of and the extent to which either party has protection from liability for costs. This is especially the case if previous 'without prejudice' offers of settlement may have been made. If it is long after the initial arbitration has been commenced that the respondent realises that a fresh notice is necessary to pursue the counterclaim, the consequences could be serious not only in terms of costs, but also in regard to the limitation period.

14.3.5 Powers of the arbitrator

The 1996 Arbitration Act significantly broadened the arbitrator's powers, than was previously the case. For example, an arbitrator may:

- Order which documents or classes of documents should be disclosed between and produced by the parties (section 34(2)(d))
- Order whether the strict rules of evidence shall apply (section 34(2)(f))
- Decide the extent to which the arbitrator should take the initiative in ascertaining the facts and the law (section 34(2)(g))
- Take legal or technical assistance or advice (section 37)
- Order security for costs (section 38)
- Give directions in relation to any property owned by or in the possession of any party to the proceedings which is the subject of the proceedings (section 38)

- Make more than one award at different times on different aspects of the matters to be determined (section 47(1))
- Award interest (sections 49(1) to 49(6))
- Make an award on costs of the arbitration between the parties (sections 61(1) and 61(2))
- Direct that the recoverable cost of the arbitration, or any part of the arbitral proceedings, is to be limited to a specified amount (sections 65(1) and 65(2)).

Figure 14.2 shows the outline of adjudication and arbitration in simple flowchart form.

14.3.6 Third party procedure

One of the perceived advantages of litigation over arbitration is that claimants can take action against several defendants at the same time, and any defendant can seek to join in another party who may have liability. This facility is not readily available in arbitration, which usually takes place only between the parties to the contract. In the IFC 98 contract an attempt was made to provide for interlocking arbitrations. In IC, the draftsman has wisely left that possibility to be covered by the CIMAR rules.

Rules 2.6 and 2.7 provide that where there are two or more related sets of proceedings on the same topic, but under different arbitration agreements, anyone who is charged with appointing an arbitrator must consider whether the same arbitrator should be appointed for both. In the absence of relevant grounds to do otherwise, the same arbitrator is to be appointed. If different appointers are involved they must consult one another. If one arbitrator is already appointed, that arbitrator must be considered for appointment to the other arbitrations.

This situation commonly occurs when there is an arbitration under the main contract and also between the contractor and a sub-contractor about the same issue, perhaps one of valuation or extension of time. It is also possible that there are two contracts between the same two parties and an issue arises in both which is essentially the same point. Usually, the same arbitrator ought to be appointed for that situation.

14.4 *Legal proceedings*

The legal proceedings option is dealt with by article 9, which simply provides that the English courts will have jurisdiction over any dispute or difference arising out of or in connection with the contract. Parties wishing to adopt this procedure will delete the arbitration option (article 8). It should be remembered that the default position has changed under this contract. If neither option is deleted, legal proceedings is the default position.

14.5 Summary

- Adjudication can be used by either party at any time, even if litigation or arbitration is in progress
- Adjudication is only temporarily binding
- It is a short and fairly 'blunt' procedure
- There is no 'appeal' against an adjudicator's decision. If referred to arbitration or the courts, the dispute will be heard again from the beginning
- The courts will not overturn an adjudicator's decision just because it is wrong
- The usual reason why a decision is not enforced is lack of jurisdiction on the part of the adjudicator
- Arbitration can be commenced by either party at any time
- It is best if the parties can agree an arbitrator
- Arbitration is to be conducted under the JCT/CIMAR Rules and the Arbitration Act 1996
- If the parties have agreed arbitration as the means of binding dispute resolution, a party can be prevented from seeking litigation by reference to section 9 of the Act
- The courts have powers equal in scope to that of an arbitrator
- Legal proceedings is the default option if no procedure is entered by the parties in the contract particulars
- All forms of dispute resolution are best treated as the utmost last resort. Even the cheapest is too expensive in overall cost, whether the claimant wins or loses.

APPENDIX A
INTERMEDIATE NAMED SUB-CONTRACTOR TENDER AND AGREEMENT (ICSUB/NAM/IT, ICSUB/NAM/T, ICSUB/NAM/A)

The provisions for named sub-contractors considered in Chapter 8 depend for their operation on the correct and full completion of these documents. The administrative burden on the architect, contractor and sub-contractor is considerable.

Invitation to tender (ICSub/NAM/IT)

The whole of this form is to be completed by the architect. It gives the proposed sub-contractor the necessary basic information on which to base its tender. Great care must be taken in its completion, because in this form and in the conditions are to be found the 'particulars' which may prevent the contractor from entering into a sub-contract with the proposed named sub-contractor (schedule 2, paragraph 2) and which are also referred to in the recitals of IC and ICD. In particular, portions of text with an asterisk * must be deleted as appropriate. For the architect to send out the form without completing more than the first page is worse than useless.

The following information is set out:

- The name and address of the tenderer, and an invitation to submit a tender by completing and returning tender form ICSub/NAM/T to the architect
- There are three alternatives: where the work is to be included in the main contract documents for pricing by the contractor, or is to be included in a provisional sum instruction, or in an architect's instruction naming the tenderer as a replacement contractor
- The 'Tender Documents' prepared on behalf of the employer which form the basis of the tender. Page 2 must then be signed and dated by the architect

- The 'Priced Documents' which the sub-contractor will be required to provide if named
- Details of the main contract Works and location, taken from the first recital of the main contract, together with any job reference
- Details of the sub-contract works
- Details of the employer, architect, quantity surveyor and contractor (if appointed)
- Relevant main contract information, including any changes in the printed conditions, whether the main contract is to be entered into as a deed or under hand, and where the main contract can be inspected (if the contractor has already been appointed)
- It is important that details of the contract particulars of the main contract are included; it fixes the named sub-contractor with notice of all the important provisions, including liquidated damages
- Access, order of Works and obligations and restrictions imposed by the employer and not covered by the main contract conditions
- Sub-contract information, including base date, dates for commencement and completion, drawings submission, attendance, interim payments, fluctuations options, insurances and dispute resolution procedures.

Tender by sub-contractor (ICSub/NAM/T)

This takes the form of an offer addressed to both the employer and the contractor. However, it will be for the contractor, not the employer, to enter into a sub-contract.

There are five pages attached to the tender in which the sub-contractor must insert relevant details, including programme, attendances, fluctuations, etc. In the tender form, it states:

> 'This tender remains open for acceptance by the Contractor within ____ weeks of the date of this Tender. . . .'

It is intended that the sub-contractor, or more likely the architect before sending out the form, will fill in the period. Despite the statement, under the general law, the proposed sub-contractor is entitled to revoke or withdraw the tender at any time prior to acceptance by the contractor. To avoid this, some other forms of tender include wording somewhat as follows:

> 'In consideration of the sum of £1 (receipt of which is hereby acknowledged), the sub-contractor agrees to keep this tender open for acceptance for a period of . . .'

This effectively makes a little contract between sub-contractor and contractor by which the contractor has paid for the benefit of having a period in which to consider the tender.

Articles of agreement (ICSub/NAM/A)

The architect is not involved in the completion of this document, which forms the sub-contract between contractor and named sub-contractor. It is in the usual form, and incorporates the sub-contract clauses ICSub/NAM/C (or 'Conditions' as they are called here) by article 1. The appropriate insertions should be made before execution.

APPENDIX B
INTERMEDIATE NAMED SUB-CONTRACT CONDITIONS (ICSUB/NAM/C)

These conditions are deemed to be incorporated in article 1 of the articles of agreement ICSub/NAM/A. They were completely revised and restructured in 2006 to match the new Intermediate Contracts IC and ICD.

The principal provisions are briefly set out below.

1.2–1.10 Interpretation, definitions, etc.

This clause states that the contract is to be read as a whole. There are detailed provisions about the priority of documents. There are provisions dealing with the law of the contract, notice provisions, third party rights, electronic communications, and how days are reckoned. The major part of the clause is taken up by a useful set of definitions, but a substantial clause deals with the effect of the final payment notice. The notice is conclusive that, where quality and standards are expressly stated to be to the satisfaction of the architect, they are to the architect's reasonable satisfaction, that effect has been given to sub-contract provisions requiring adjustment of the sub-contract sum, that extensions of time have been given, that reimbursement has been made of loss and/or expense to the sub-contractor and to the contractor. It is not conclusive if proceedings including adjudication commenced before or within 10 days of the contractor's notice to the sub-contractor of the amount or in the case of accidental inclusion or deduction of items or of arithmetical errors.

2.1 and 2.2 Sub-contractor's obligations – carrying out and completion of sub-contract works

The sub-contractor must carry out and complete the works, including NAM designed works, in a proper and workmanlike manner, in accordance with the health and safety plan, and in accordance with the dates and periods stipulated in the sub-contract and reasonably in accordance with progress of the main contract Works.

2.3 *Materials, goods and workmanship*

Where approval of workmanship or materials is to be to the satisfaction of the architect, they are to be so.

2.4 *Sub-contractor's liability under incorporated provisions of the main contract*

The sub-contractor is to observe and comply with all main contract provisions as far as they relate to the sub-contract, and the sub-contractor is to indemnify the contractor against breach, act or omission of the main contract provisions insofar as they relate to the sub-contract and against any claim resulting from the sub-contractor's negligence. But the sub-contractor has no liability in respect of negligence by the employer, contractor, other sub-contractors, or their servants or agents.

2.5–2.6 *Construction information*

The contractor must supply two copies of drawings relevant to the sub-contract works which are supplied to the architect. The sub-contractor must provide the contractor with other information necessary for the construction of the sub-contract work, including design information where appropriate.

The information must be given in due time to the relevant party. Design information must be provided to the architect and to the contractor by the sub-contractor.

2.7–2.9 *Bills of quantities and directions on errors*

These must be prepared in accordance with SMM. Departures from the method of preparation of bills of quantities are to be corrected by the contractor's directions. Under certain conditions, if directions vary the work, they are to be valued as variations.

2.10 *Divergences from statutory requirements*

This deals with the treatment of divergences, which are generally to be corrected and treated as variations.

2.11 *Unfixed materials*

Unfixed materials must not be removed unless the contractor consents. Provisions are included which are intended to ensure that, if the value of any materials has been included in an architect's certificate, the goods become the property of the employer or, if the contractor has paid for them, of the contractor.

2.12–2.13 *Adjustment of period for completion*

If it is reasonably apparent that progress is or will be delayed, the sub-contractor must give notice of the delay, specifying the cause. If, for certain causes, the works are likely to be delayed beyond the period fixed for completion, the contractor must make in writing a fair and reasonable extension of time as soon as it is able. The causes are: variations; compliance with certain contractor's directions; deferment of possession; inaccurate approximate quantities; valid suspension by the sub-contractor under clause 4.13; valid suspension by the main contractor under the main contract clause 4.11; employer's default; contractor's default; statutory undertakers' work; exceptionally adverse weather conditions; loss or damage by specified perils; civil commotion or terrorism; strike, etc.; exercise of statutory power; *force majeure*. There is provision for extension of time if certain defaults or events occur after the expiry of the periods for completion. At any time up to 16 weeks after practical completion of the sub-contract works the contractor may make an extension of time as a result of a review of previous decisions or otherwise, provided that previous extensions are not reduced. The sub-contractor must use its best endeavours to prevent delay and do all reasonably required by the contractor to proceed with the works. It must also provide such further information as the contractor requires.

2.14 *Practical completion*

The sub-contractor must notify the contractor in writing when it considers that practical completion of the sub-contract works has been achieved, and if the contractor does not dissent within 14 days, practical completion is deemed to have taken place on the date notified. The contractor must give reasons if it dissents, and the date will be as notified by the contractor, or the adjudicator or arbitrator must decide when practical completion has occurred. If there is no agreement or decision as a result of the dispute resolution procedure, practical completion will be when the architect certifies practical completion of the main contract works.

2.15 *Failure of sub-contractor to complete on time*

If the sub-contractor fails to complete on time, the contractor must give it a notice to that effect within a reasonable time. Then the sub-contractor must pay or allow loss and/or expense caused to the contractor by the failure.

2.16 *Sub-contractor's liability for defects*

The sub-contractor is liable to make good, at no cost to the contractor, defects, shrinkages and other faults in respect of sub-contract works.

2.17 *Deductions under the main contract*

If the architect instructs that certain defects are not to be made good or setting-out errors not corrected but a deduction is to be made from the contract sum under clause 2.9 or 2.30 of the main contract, the contractor will pass on the instruction and deduction to the sub-contractor insofar as it affects the sub-contract works.

2.18 *As-built drawings*

These are to be supplied by the sub-contractor before practical completion where NAM designed works are involved.

2.19 Copyright

Copyright is vested in the sub-contractor, but the contractor has irrevocable licence with power to pass a sub-licence to the employer.

3.1–3.3 Assignment – sub-letting

The sub-contractor is not allowed to assign the sub-contract or sub-let any portion of the Works without the consent of the contractor. There is provision for interest on late payment under the sub-sub-contract.

3.4–3.6 Directions of the contractor

The contractor may issue reasonable directions in writing. The architect's written instructions affecting the sub-contract works are to be taken as from the contractor. Variations will not vitiate the sub-contract. The sub-contractor must comply forthwith unless the variation relates to 'obligations or restrictions' in regard to access etc., when it may make reasonable objection. The contractor may issue a seven-day notice of compliance and, if this is ignored, can employ others, pay them and deduct the money from monies due to the sub-contractor.

3.7 Person-in-charge

The sub-contractor must ensure a person-in-charge is on site at all times during execution of sub-contract works.

3.8 Right of access for contractor and the architect

The contractor and the architect and their representatives must be given access to sub-contract work in preparation.

3.9–3.10 Inspections, tests and work not in accordance with the sub-contract

The contractor may issue directions for opening up and testing, but if the work is in accordance with the sub-contract, the cost must be added to the sub-contract sum. If any work or materials fail, the contractor has the power to ask that similar work be opened up and tested, provided that it uses the power reasonably. Reasonableness may be tested by adjudication.

3.11 *Removal of work*

The contractor may issue directions about the removal of work not in accordance with the contract.

3.12 *Non-compliant work by others*

The sub-contractor must comply with directions of the contractor regarding the taking down and re-erection of work resulting from activities of other sub-contractors complying with contractor's directions, and the sub-contractor shall be entitled to be paid therefor.

3.13 *Sub-contractor's indemnity*

The sub-contractor must indemnify the contractor against losses incurred by the contractor as a result of the sub-contractor complying with clauses 3.9–3.11 or as a result of the operation of clause 3.12.

3.14 *Attendance*

This clause sets out the specific items of attendance which the sub-contractor may expect free of charge from the contractor. The sub-contractor's responsibilities are detailed, together with its rights in regard to erected scaffolding. There are prohibitions against wrongful use of the other's equipment and against infringement of Acts of Parliament, regulations, etc. This clause is without prejudice to the parties' rights to carry out their respective statutory or contractual duties.

3.15 *Temporary buildings*

This provides that the sub-contractor must erect all its own temporary buildings as the contractor directs.

3.16 Site clearance

This provides for the sub-contractor to clear away rubbish.

3.17–3.18 Health and safety and the CDM Regulations

The sub-contractor must comply with all health and safety legislation and the reasonable directions of the contractor in that regard. Clause 3.18 applies where the main contract particulars so state. It sets out the obligations of the contractor and sub-contractor to comply with the CDM Regulations by way of providing information and complying with the reasonable requirements of the principal contractor.

3.19–3.20 Suspension of main contract

The contractor must copy the sub-contractor with any notice to the employer threatening suspension. The contractor may issue directions to the sub-contractor to cease work, and the sub-contractor must resume work if so directed.

3.21 Strikes

If the works are affected by strikes or lockouts, neither party may make any claim on the other therefor; the contractor must try to keep the Works available for the sub-contractor, and the sub-contractor must try to proceed with its own work. This clause does not affect any other rights of either party under the sub-contract.

3.22 Benefits under main contract

The contractor must obtain any lawful benefits of the main contract for the sub-contractor, provided that the sub-contractor so requests and is prepared to pay any costs involved.

3.23 Certificates under the main contract

If the sub-contractor so requests, the contractor must notify the sub-contractor of the dates of all such certificates.

4.1–4.4 Quality and quantity of work and final sub-contract sum

Where bills of quantities are included, the quality and quantity of work are to be as set out therein and in the NAM design requirements. If there are no bills of quantities, but quantities are contained in the numbered documents, they will control the quality and quantity of work. Otherwise, the sub-contract documents taken together will determine the quality and quantity of the work, with the contract drawings prevailing. Where there is an adjustment basis, the sub-contract sum may be altered only under the express provisions of the contract (essentially by adding or omitting). Where there is a remeasurement basis, the whole of the sub-contract works will be subject to remeasurement. Clauses 4.3 and 4.4 detail the provisions for calculating the final sub-contract sum on adjustment and remeasurement bases respectively.

4.5 Taking adjustments into account

Provisions for additions or deductions in the conditions are to be taken into account in the next interim payment following ascertainment.

4.6 Calculation of the final sub-contract sum

All documents reasonably required for calculating the final sub-contract sum must be sent to the contractor before or not later than four months after practical completion of the sub-contract works. The contractor has a further eight months in which to send the sub-contractor a statement of the calculation of the final sub-contract sum.

4.7 Value added tax

This is providing for recovery of tax. It also provides that the sub-contract sum is to be exclusive of tax.

4.8 Construction Industry Scheme (CIS)

This makes the contractor's obligation to pay subject to the CIS.

4.9–4.10 Interim payments

Interim payments will be due on the same date each month, the first payment becoming due as stated in article 3 or one month at latest from commencement of sub-contract works on site. The contractor has 21 days to pay from the due date. The clause details which amounts are to be included and which amounts are to be subject to retention. Unfixed materials for incorporation must not be removed without the contractor's consent.

4.11 Listed items

This clause provides for payment for materials off-site subject to certain conditions.

4.12 Interim payment notices

This clause provides for the giving of notices to comply with the Housing Grants, Construction and Regeneration Act 1996. This includes notices of the amount proposed to be paid, amount of withholding, and interest on money which the contractor has wrongly fully failed to pay.

4.13 Suspension rights

An important provision allows the sub-contractor to suspend performance of obligations if the main contractor fails to pay as provided seven days after receipt of the sub-contractor's notice.

4.14 Final payment

Final payment is due not later than seven days after the architect's final certificate under the main contract. Payment must be made within 28 days of the due date. Notices similar to the ones set out in clause 4.12 are to be sent to the sub-contractor, and interest is payable on late payments.

4.15 *Fluctuations options*

This refers to option A or C in schedule 2 of the sub-contract.

4.16–4.17 *Loss and/or expense*

The agreed amount must be added to the sub-contract sum if the sub-contractor makes a written application to the contractor within a reasonable time of its becoming apparent that it is likely to incur direct loss and/or expense because the sub-contract works are being affected by any of the following: variations; certain directions of the contractor including architect's instructions; valid suspension by the sub-contractor under clause 4.13; valid suspension by the contractor under the main contract clause 4.11; inaccurate approximate quantities; employer's defaults; contractor's defaults.

4.18 *Contractor's reimbursement*

If regular progress is materially affected by the sub-contractor's own default, and the contractor makes a written application within a reasonable time thereafter, the agreed amount of direct loss and/or expense may be deducted from monies due from the contractor to the sub-contractor, or it may be recoverable by the contractor as a debt. The contractor must provide such details as the sub-contractor requires. These provisions are without prejudice to the other rights and remedies possessed by the parties.

4.19 *Reservation of rights*

This clause reserves all the contractor's and the sub-contractor's rights and remedies.

5.1 *Definition of variations*

Generally, the alteration or modification of the design, quality or quantity of work.

5.2–5.3 Valuation of variations on an adjustment basis

The contractor and sub-contractor may agree the amount to be added to, or deducted from, the sub-contract sum in connection with a variation before the variation is carried out. Otherwise, omissions must be valued at the relevant prices in the bills of quantities or other sub-contract documents. Work similar in character to that in the bills of quantities is to be consistently valued, making due allowance for any changes. If there is no work of similar character, if the work to be valued is not added, omitted or substituted, or if it is not reasonable to do otherwise, the work must be valued at fair rates and prices. Approximate quantities which are reasonably accurate must be valued on the basis of the rate for that approximate quantity, otherwise a fair allowance must be made for any difference in quantity.

5.4 Valuation of variations on a remeasurement basis

This work is to be measured and valued. Work similar in character to that in the bills of quantities is to be consistently valued, making due allowance for any changes in conditions or quantity. If there is no work of similar character, if the work to be valued is not added, omitted or substituted, or if it is not reasonable to do otherwise, the work must be valued at fair rates and prices.

5.5–5.6 General rules and daywork

Measurement must be in accordance with the principles governing the preparation of the bills of quantities; allowance must be made for lump sum adjustments and for adjustment of preliminary items. The definition of prime cost in relation to daywork is similar to that in IC.

5.7 Change of condition

Where variations to work change the condition under which other work is executed, such other work is to be treated as if subject to a variation direction.

5.8 Additional provisions

A fair valuation must be made if valuation cannot be done in any other way. No allowance must be made for loss and/or expense reimbursable under any other provision.

6.1 Insurance definitions

Definitions of excepted risks, joint names policy, specified perils, terminal date and terrorism cover.

6.2–6.3 Liability of sub-contractor – injury to persons and property

The sub-contractor is liable for, and must indemnify the contractor against, loss etc. arising from personal injury or death in connection with the sub-contract works unless and to the extent it is caused by the negligence of the contractor or of the employer. The sub-contractor must generally indemnify the contractor against loss etc. arising from damage to property caused by the carrying out of the sub-contract works unless and to the extent it is caused by the negligence of the contractor or employer.

6.4–6.5 Insurance – sub-contractor

Subject to certain exceptions, the sub-contractor must maintain insurance to cover the liabilities in clauses 6.2 and 6.3.

6.6–6.7 Loss or damage by specified perils to the works and materials on site

Prior to the commencement of the sub-contract works the contractor must ensure that either the sub-contractor is recognised as an insured or the insurers waive rights of subrogation. The sub-contractor is not responsible for the cost of restoration of the sub-contract works if the loss or damage is due to a specified peril or negligence by the contractor or if the employer of the contractor does not make a claim. The sub-contractor is responsible insofar as the loss is due to perils other than specified perils. When reasonably required to do so, the sub-contractor must produce documentary evidence (policies and premium receipts) that it has taken out the appropriate insurance. If the sub-contractor fails to insure, the contractor may itself insure and deduct the premium money from monies due to the sub-contractor. The sub-contractor must give notice on discovering the loss or damage, and commence restoration and repair work as directed by the contractor. Where the sub-contractor is not responsible, compliance with the contractor's directions is to be treated as a variation. The occurrence of loss or damage is to be ignored in computing the amounts payable to the sub-contractor under the sub-contract.

6.8 *Terrorism cover – non-availability options*

Provisions if insurers notify employer or contractor that terrorism cover will cease.

6.9 *Sub-contractor's responsibility for its own plant*

Property of the sub-contractor on site and not for incorporation is at the sub-contractor's risk as regards loss or damage not caused by the negligence of the contractor.

6.10–6.13 *Joint Fire Code*

Putting the Joint Fire Code into effect.

7.1–7.3 *Termination general*

Insolvency is defined, notices not to be given unreasonably or vexatiously and delivery by specified means. The parties' rights are preserved.

7.4–7.7 *Termination of the employment of the sub-contractor by the contractor*

The contractor may terminate if:

- The sub-contractor substantially suspends the work or design without reasonable cause
- The sub-contractor does not proceed with the sub-contract Works or the design regularly and diligently
- The sub-contractor neglects to comply with the contractor's notice to remove defective works and the main contract Works are materially affected thereby, or fails its obligations in respect of remedial work
- The sub-contractor contravenes the provisions for assignment and sub-letting
- The sub-contractor fails to comply with the CDM Regulations.

This is provided that the sub-contractor has failed to rectify its default 10 days after the contractor's written notice. If the sub-contractor becomes insolvent or is guilty of corruption, the contractor can terminate forthwith.

Detailed provisions are included regarding the procedure following termination. Items covered are the removal of equipment from the Works, employment of others to carry out the sub-contract Works, and payment.

7.8–7.11 Termination of employment under the sub-contract by the sub-contractor

The sub-contractor may terminate if the contractor substantially suspends the main contract Works without reasonable cause, or seriously affects the sub-contractor's work by failing to proceed with its own work without reasonable cause, or fails to pay as required by the sub-contract or fails to comply with the CDM Regulations.

This is provided that the contractor has failed to rectify its default 10 days after receipt of the sub-contractor's written notice. If the contractor subsequently repeats the same default, the sub-contractor may terminate forthwith. Termination is possible in the case of contractor insolvency, but there are certain provisions to be satisfied.

The sub-contractor's employment terminates if the contractor's employment is terminated under the main contract.

Detailed provisions are included regarding the procedure to be followed after termination. Removal of equipment from site and payment are covered. The sub-contractor is entitled to recover direct loss and/or expense.

8.1 Mediation

Disputes may be resolved by mediation if the parties so agree.

8.2 Adjudication

The procedure will be under the Scheme for Construction Contracts (England and Wales) Regulations 1998 (or the equivalent Northern Ireland or Scottish Schemes if the Works are in either of those jurisdictions).

8.3–8.8 Arbitration

The JCT 2005 edition of the Construction Industry Model Arbitration Rules (CIMAR) and the Arbitration Act 1996 apply.

APPENDIX C

INTERMEDIATE NAMED SUB-CONTRACTOR/EMPLOYER AGREEMENT (ICSUB/NAM/E)

Under IC and ICD, for certain defaults of a named sub-contractor, the employer can suffer loss or damage and have no contractual remedy against the main contractor. This is because of the terms of paragraph 11.1 of schedule 2, which exempts the main contractor from liability to the employer in respect of a named sub-contractor's failure to exercise reasonable care and skill in any of the following:

- The design of the sub-contract works so far as the named sub-contractor has designed or will design them
- The selection of the kinds of materials and goods for the sub-contract works so far as such goods etc. have been or will be selected by the named sub-contractor
- The satisfaction of any performance specification or requirement relating to the sub-contract works.

Almost inevitably there will be a design or related element in a named sub-contractor's work. Paragraph 11.1 reverses the general rule of law, which is that the main contractor is responsible for sub-contractors' defaults of design, fabrication or otherwise.

The specified areas are ones where the employer needs further protection, which is also necessary if the sub-contractor fails to provide information to the architect and causes late instructions to be given to the contractor, and as a result the contractor has a valid claim for extension of time and extra cost.

These are the main areas which ICSub/NAM/E seeks to cover, the device adopted being a direct contract between the employer and the named person, which is intended to be completed contemporaneously with the submission of a tender in Form ICSub/NAM. ICSub/NAM/E is similar to the previous Form ESA/1. However, ESA/1 was prepared not by the Joint Contracts Tribunal but by the Royal Institute of British Architects and the Committee of Associations of Specialist Engineering Contractors, and it was a fairly short form.

The new form is somewhat longer, and care will be needed in its completion. It includes comprehensive and good guidance notes on its correct completion. It is essential that these notes are read carefully before there is any attempt to complete the form. The functions of the agreement are:

- To provide a direct line of redress for the employer regarding design work by the sub-contractor
- To provide a vehicle for the execution of design or other work prior to the sub-contract being executed and to enable payment to be made for such work if the sub-contract is not executed
- To enable the sub-contractor to provide warranties to purchasers, tenants and/or funders
- To provide for professional indemnity insurance by the named sub-contractor if design work is done.

In structure, the form consists of articles and recitals and a set of clauses:

(1) Supply of designs and information
(2) Materials, goods and workmanship
(3) Approximate estimate – tender
(4) Execution of named sub-contract
(5) Payment of costs and expenses
(6) Collateral warranties
(7) Use of design information
(8) Termination under main contract or named sub-contract
(9) Notices under clauses 6 and 8
(10) Assignment
(11) References to legislation
(12) Contracts (Rights of Third Parties) Act 1999
(13) Disputes, law and jurisdiction.

These are followed by the attestation and schedules:

(1) Information requirements
(2) Procurement and/or fabrication of materials and goods
(3) Collateral warranties

Table of Cases

Clause Number Index to Text

Subject Index